A GUIDED TOUR OF
DIFFERENTIAL EQUATIONS
USING COMPUTER TECHNOLOGY

I hear, I forget

I see, I remember

I do, I understand

Anonymous

ALEXANDRA SKIDMORE, PH. D.
Emory & Henry College

MARGIE HALE, PH. D.
Stetson University

PRENTICE HALL, Upper Saddle River, NJ 07458

Executive Editor: George Lobell
Production Editor: Dawn Blayer
Supplement Cover Designer: PM Workshop Inc.
Special Projects Manager: Barbara A. Murray
Supplement Cover Manager: Paul Gourhan
Supplement Editor: Audra Walsh
Manufacturing Buyer: Alan Fischer

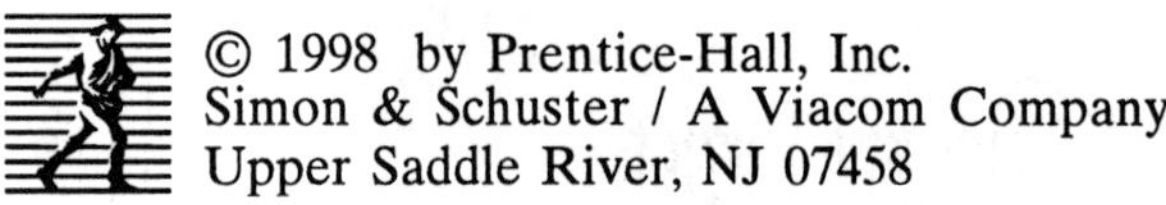

Printed in the United States of America

10 9 8 7 6 5 4 3 2 1

ISBN 0-13-592767-6

Prentice-Hall International (UK) Limited, *London*
Prentice-Hall of Australia Pty. Limited, *Sydney*
Prentice-Hall Canada, Inc., *London*
Prentice-Hall Hispanoamericana, S.A., *Mexico*
Prentice-Hall of India Private Limited, *New Delhi*
Prentice-Hall of Japan, Inc., *Tokyo*
Simon & Schuster Asia Pte. Ltd., *Singapore*
Editora Prentice-Hall do Brazil, Ltda., *Rio de Janeiro*

A GUIDED TOUR OF
DIFFERENTIAL EQUATIONS
USING COMPUTER TECHNOLOGY

A GUIDED TOUR OF DIFFERENTIAL EQUATIONS
USING COMPUTER TECHNOLOGY

CONTENTS

This book is dedicated to

my husband, friend, and colleague,

James A. Warden

and to the memory of my father,

Victor J. Samanich

. . . Sandy Skidmore

We are very grateful to the following friends and colleagues: all of our patient students, who were marvelous guinea pigs; Beverly West for her enthusiastic support and encouragement; Jim Wahab for his faith in the project; Dan Dubois of Maple for his ever prompt, courteous, and helpful suggestions; our understanding and most helpful editors, George Lobell and Audra Walsh; and all the staff of the Computer Center at Emory & Henry College.

Margie Hale wishes to acknowledge with gratitude the support she received from Ithaca College in developing these materials. This support took the form of encouragement from her colleagues there and a summer grant through the National Science Foundation administered by the Mathematics Department.

Sandy Skidmore deeply appreciates the patience and encouragement of her mathematics colleagues at Emory & Henry College, and a Summer Curriculum Development Grant she received from the College.

i

TO THE INSTRUCTOR

This collection of informal Lessons is intended to be used by students as guided inquiries to lead them through the basic ideas of differential equations and to help them develop an intuitive understanding of the subject. The Lessons should be used as a supplement to a text, since they are by no means exhaustive of a topic, nor are they intended to be.

We recommend that students work through a Lesson before reading the formal presentation on that topic in their text, because doing a Lesson should make the reading that much easier. Or they might do a Lesson after having heard a formal lecture, in order to reinforce the basic ideas presented. Most of the Lessons will require about an hour of the student's time; a few are a little longer.

Although the topics are covered in a "traditional" order, we have treated them in a more contemporary fashion, by providing reasons and motivation wherever possible, and by including qualitative behavior studies. In developing mathematical maturity, students must learn to take risks, be given a chance to recognize patterns, try to generalize from examples. This takes practice. One goal of this book is to give them some practice.

The Lessons are intended to serve several purposes:

- Since students usually have difficulty with "word problems", several of the Lessons lead them through the modeling process step by careful step (see the Lessons on mixing problems, population growth, the mass-spring system, and the simple pendulum.)

- In other Lessons, they are guided through derivations of various solution methods, always using concrete examples. They are often asked to anticipate the next step in the development. We believe that people learn by doing and questioning, be it walking, talking, or doing mathematics.

- Some Tasks in each Lesson are designed to solidify what they've just learned, while others guide them to make observations and "educated guesses" about the nature of the problem at hand.

- The Lessons make good and frequent use of a CAS to help them on their way toward understanding and "seeing" DEs. Examples and syntax for Maple V Release 4 and Mathematica, Versions 2.2.2 and 3, are included in the Appendix. As a byproduct, students also get to see how today's software might break down, so that developing "by hand" techniques to check the computer can pay off.

A word of caution: Drawing phase portraits for systems is slow and non-dynamic in Maple and Mathematica, besides using a lot of memory. Faster and more gratifying tools are MacMath or PHASER. Nevertheless, we've included only commands for Maple and Mathematica in the Appendix. We've also included some web site addresses.

We have used versions of these Lessons for the past several years in our sophomore-junior level DE courses at small liberal arts colleges, using the method of cooperative learning. The students seem to enjoy them, and we're pleased with the results.

Our hope is that students will have an easier time learning differential equations, instructors will have an easier time teaching the course, and everyone should have some fun!

We're always open for questions, comments, and corrections. Mail either one of us at:

Sandy Skidmore: askidmor@ehc.edu
Margie Hale: mhale@stetson.edu

I hear, I forget
I see, I remember
I do, I understand.
ANONYMOUS

TO THE STUDENT

If you're one of the many students who finds reading mathematics difficult, you might like these Lessons. One of our goals as teachers is to help you gain a better understanding of the basic ideas of differential equations, so we've tried to write the Lessons in a relaxed style, unlike the formal style of most texts. You should use the Lessons as an aid to your assigned text readings and homework. We suggest that you do a Lesson on a topic before you read about it in your text, because the Lesson will provide a short introduction and give you an overview.

The Lessons use examples to explain ideas, and they lead you through solution processes step by step, with many hints and answers right there. We believe that people learn by doing, be it walking, talking, or doing math. Most of the Lessons will take about an hour of your time; a few are a little longer. Our students have found doing them to be a very good investment of their time. Working in groups can be especially helpful.

Another goal is to show you how a computer algebra system (CAS), either Maple or Mathematica, can be used to help you do the necessary computations, which are often tiresome when done "by hand". (But don't think that you don't have to learn to do this stuff "by hand"; you always have to second-guess a computer's output, so you have to have a hunch about what to expect ahead of time.) Whenever a Lesson asks you to "use your CAS" to do something, there will be a corresponding Appendix lesson that gives you the syntax, and you should type and execute it. (We assume that you already know how to do some basic things with your CAS, like a little algebra, plotting curves, differentiation and integration.) Of real use is your CAS's help feature; you should use it frequently.

Both the input and output of the Maple sessions are included. Because Mathematica's output is so similar to Maple's, we've included its output only when we thought it was needed for clarity. If you're a Mathematica user, reading the Maple output will help you see what to expect from Mathematica.

Because we've tried to be informal in our mathematical presentation, we've omitted all the necessary hypotheses on continuity and differentiability of functions, since most of the ones we use are our old friends from calculus. However, you should consult your text for the detailed statements of theorems.

To help you in your adventures, we've included a list of web sites (current as of June 1997) that you might wish to investigate.

We're always open for questions, comments, and corrections. Mail either one of us at:

Sandy Skidmore: askidmor@ehc.edu
Margie Hale   : mhale@stetson.edu

THE LESSONS

- **A useful way to think about functions and their derivatives**

Thinking about the relationship between a function and its derivative, in words, is often the first step in formulating a differential equation which describes a real situation. This lesson will give you some practice. Save your work for use in later lessons.

Task 1a. Find the derivative of each of the following functions. Check your answers using your computer algebra system (known as a CAS). The appropriate syntax for either Maple or Mathematica is in the Appendix.

FUNCTION	DERIVATIVE
1. $y = e^t$	1.
2. $y = e^{3t}$	2.
3. $y = e^{(-0.1)t}$	3.
4. $y = mt$ (m constant)	4.
5. $y = kt^3$ (k constant)	5.
6. $y = \sqrt{2t + 1}$	6.

Task 1b. Now match each statement below to the function or functions above which satisfy the statement.

_______A. The derivative of the function is -0.1 times the function.

_______B. The rate of change of the function is constant.

_______C. The derivative of the function is three times the function over t .

_______D. The derivative of the function is the function.

_______E. The rate of change of the function is proportional to the function.

_______F. The derivative of the function is the reciprocal of the function.

_______G. The derivative of the function is the function over t .

________H. The derivative of the function is three times the function.

________I. The function changes at a rate that is inversely proportional to the function.

________J. The derivative of the function is always positive.

________K. The function is always increasing.

________L. The rate of change of the function is always negative.

________M. The function is always decreasing.

Task 2. Using the functions given in Task 1a., match the following differential equations to the functions, that is, figure out which function makes an equation true. Your verbal descriptions of the derivatives from Task 1b. will be helpful.

————I. $\dfrac{dy}{dt} = 3y$ ————IV. $\dfrac{dy}{dt} = y$

————II. $\dfrac{dy}{dt} = \dfrac{3y}{t}$ ————V. $\dfrac{dy}{dt} = \dfrac{1}{y}$

————III. $\dfrac{dy}{dt} = -0.1y$ ————VI. $\dfrac{dy}{dt} = \dfrac{y}{t}$

The six equations in Task 2 above are called ***first-order differential equations*** because the highest derivative that appears is the first derivative, *dy/dt*. Each of the six functions of Task 1 is a ***solution*** of one of the differential equations, because the function makes the differential equation "true" for some t-interval of real numbers, the domain of the solution.

- **Equations with variables separable**

One of the easiest kinds of first-order differential equations to solve is one whose variables can be "separated" using algebra, in the sense that all the y's will appear on one side of the equation and all the t's will appear on the other side. Here's an example:

Given the differential equation

$$\frac{dy}{dt} = \frac{t}{y}$$

think of dy and dt as separate entities, and do some algebra to rearrange the equation as

$$y \, dy = t \, dt.$$

Now the variables are "separated" because all the $y's$ are on one side of the equation and all the $t's$ are on the other side. Also, the dy and the dt are in the numerators. This is important because the next step is integration. Integrate both sides to get

$$\int y \, dy = \int t \, dt \, ,$$

which becomes

$$\frac{y^2}{2} = \frac{t^2}{2} + C \, ,$$

where both integration constants have been lumped together as C. This last equation can be rearranged to give the solution curves

$$y^2 - t^2 = C_1 \, , \quad (where \ C_1 = 2C) \, .$$

These curves are hyperbolas, which are <u>not</u> functions of t; but we can solve for y to get the functions:

$$y(t) = \pm \sqrt{t^2 + C_1} \, .$$

This result is called a ***one-parameter family*** of solutions, or ***a general solution***. The constant C_1 is called a ***parameter***. Each choice of the constant C_1, and a choice of either the plus or the minus sign, will yield a solution of the differential equation. What are the domains of these functions?

In general, when you solve a first-order differential equation, by whatever method, you'll end up with infinitely many solutions, because of the integration constant. (There are some minor exceptions to this rule which you need not worry about.)

Summary of the method: To solve a separable differential equation given in the form

$$\frac{dy}{dt} = f(t) \cdot g(y) ,$$

do the following:

1) Use algebra to get all the y's and the dy on one side of the equation and all the t's and the dt on the other side, making sure that dy and dt end up in the numerators;

2) Integrate both sides of the resulting equation, and don't forget the constants of integration;

3) Solve the resulting equation for y as a function of t, if you can (this won't always be possible). You should end up with an entire family of solutions; each choice of the constant will give a different solution.

Task 1. Solve each of the following differential equations by the method of separation of variables.

1. $\dfrac{dy}{dt} = y$

2. $\dfrac{dy}{dt} = -2y$

3. $\dfrac{dy}{dt} = \dfrac{y}{t}$

4. $\dfrac{dy}{dt} = \dfrac{t}{y}$

5. $\dfrac{dy}{dt} = \dfrac{t^2}{y^2}$

6. $\dfrac{dy}{dt} = \dfrac{1}{y}$

7. $\dfrac{dy}{dt} = y - y^2$

[Hint for #7: To do the integration, use the method of partial fractions, an integral table, or the integration command in your CAS.]

Task 2. Determine which of the following equations is separable. Can you give informal reasons why some of them cannot be separated?

1. $(2t - 1)(y - 1)^2 dy + (t - t^2)(1 + y)dy = 0$

2. $\dfrac{dy}{dt} = e^{t-y}$

3. $(\sin y)dt + (y - \cos t)dy = 0$

4. $\dfrac{dy}{dt} = \dfrac{t - y}{t + y}$

Task 3. In Step 2. of the above Summary of the Method, one side of the equation appears to be integrated with respect to the variable y, while the other side is integrated with respect to t. This doesn't make sense, since both sides of an equation represent the same thing. Figure out what's going on.

- **Verifying solutions by hand**

Below are the differential equations you solved in Task 1 of Lesson 2, and their solution families. Check to see that your solutions agree with these.

	EQUATION	*SOLUTION*
1.	$\dfrac{dy}{dt} = y$	$y = ce^{t}$
2.	$\dfrac{dy}{dt} = -2y$	$y = ce^{-2t}$
3.	$\dfrac{dy}{dt} = \dfrac{y}{t}$	$y = ct \quad (note \ t \neq 0)$
4.	$\dfrac{dy}{dt} = \dfrac{t}{y}$	$y = \pm \sqrt{t^2 + C}$
5.	$\dfrac{dy}{dt} = \dfrac{t^2}{y^2}$	$y = \sqrt[3]{t^3 + C}$
6.	$\dfrac{dy}{dt} = \dfrac{1}{y}$	$y = \pm \sqrt{2t + C}$
7.	$\dfrac{dy}{dt} = y - y^2$	$y = \dfrac{e^{t}}{e^{t} + C} \quad or \ y = 0$

You can always check that a function is a solution of a differential equation by differentiating the function and "plugging it in." For example, let's check the answer in Problem 5 above. Given that the general solution is

$$y = \sqrt[3]{t^3 + C} \, ,$$

the left side of the differential equation becomes

$$\frac{dy}{dt} = \frac{1}{3} \left(t^3 + C\right)^{-\frac{2}{3}} 3t^2 = \frac{t^2}{\left(t^3 + C\right)^{\frac{2}{3}}}$$

while the right side becomes

$$\frac{t^2}{y^2} = \frac{t^2}{\left(\sqrt[3]{t^3 + C}\right)^2} = \frac{t^2}{\left(t^3 + C\right)^{\frac{2}{3}}} \, .$$

Since both sides are identical for all real t except one, each member of the family is a solution. If you don't get a "match" for all t on some interval, you don't have a solution.

Task 1. Verify that the other families of functions in 1 through 7 of the above list are solutions of their respective differential equations. (You might want to use your CAS to do the algebra in some of them).

- **Using your CAS to find analytic solutions**

    You did a lot of work to solve these seven DEs, and you got what are called ***closed-form*** or ***analytic*** solutions, which means the solutions can be expressed using familiar algebraic formulas. This is not always the case! In fact, most useful DEs don't have closed-form solutions, and you have to resort to numerical approximations, which you'll learn about in a later lesson. For the time being, you'll learn how to use your CAS to get analytic solutions, when possible.

Task 2. Use your CAS to solve the seven DEs above. See the Appendix for the syntax. The computer's answers might look different from yours, so be prepared to check that they're algebraically equivalent.

• Using your CAS to plot families of solutions

To get further information about how solutions behave, it is useful to plot several solution curves on the same axes. This process is part of what's called studying the *qualitative* behavior of the differential equation, as opposed to the *quantitative* behavior. Here are some questions you might ask when you do this: Are the solutions bounded? oscillatory? increasing? decreasing? Are there maxima? minima? Do some solutions exhibit different behavior from all the others?

For example, the solutions of differential equation 1 of Lesson 3 corresponding to $C = 0$, ± 1, ± 2 are plotted here. The equation is $dy/dt = y$ and the solutions are $y = Ce^{t}$.

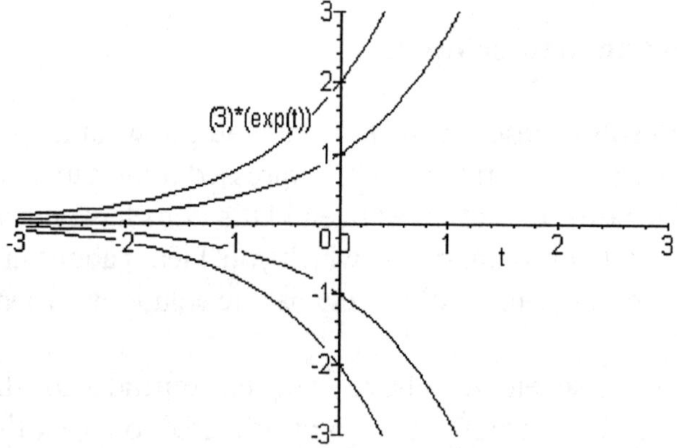

You might guess from looking at these five curves that the rest of the solutions are increasing when $C > 0$, decreasing when $C < 0$, and the solution for $C = 0$ is the horizontal line $y = 0$ (this solution is called ***the trivial solution***). Also notice that the solutions don't seem to intersect; this is an important observation, and it's almost always true.

Task 1. Use your CAS to plot, on the same axes, the solutions of the differential equations 2 through 7 of Lesson 3 corresponding to $C = 0$, ± 1, ± 2. (The syntax you'll need is in the Appendix.) Do some solutions appear to intersect? If so, where?

Caution: Sometimes computer-generated graphs are incomplete, or even wrong! Software is not infallible. So you have to develop "a feel" for what to expect, and always check the computer's answers "by hand", by plugging in some points.

- **Solving initial value problems**

Very often, especially in applications, you'll need to find a particular solution that goes through a given point of the (t,y)-plane. For example, if you want the solution of the equation

$$\frac{dy}{dt} = y$$

that goes through the point (1,2), then you are asking for the solution which satisfies the *initial condition* $y(1) = 2$. You hope there's only one such solution, and the fact that the solution curves don't appear to intersect makes you believe there is only one (there is). To find it, recall that the solution family is $y = Ce^t$; so set $t = 1$ and $y = 2$ and solve for C, as follows:

$$2 = Ce^1 = Ce,$$

so $$C = 2/e \approx 0.736 .$$

Thus the solution you're looking for is $y = 0.736e^t$.

The pair of equations discussed above, namely,

$$\frac{dy}{dt} = y , \qquad y(1) = 2$$

is called an *initial value problem*, or **IVP**. In general, *a first-order initial value problem* is a differential equation $dy/dt = f(t,y)$ along with an initial condition $y(t_0) = y_0$ that must be satisfied by the solution.

Task 2. Listed below are the seven differential equations you solved in Lesson 3, along with their general solutions. Also given is an initial condition for each equation. Look at the graphs you plotted in Task 1 and pick out or draw the solution that satisfies the given initial condition.

EQUATION	*SOLUTION*	*INITIAL CONDITION*
1. $\dfrac{dy}{dt} = y$	$y = ce^{t}$	$y(0) = -1$
2. $\dfrac{dy}{dt} = -2y$	$y = ce^{-2t}$	$y(0) = 0.5$
3. $\dfrac{dy}{dt} = \dfrac{y}{t}$	$y = ct$ (*note* $t \neq 0$)	$y(-1) = 2$

4. $\dfrac{dy}{dt} = \dfrac{t}{y}$ $\qquad\qquad$ $y = \pm\sqrt{t^2 + C}$ $\qquad\qquad$ $y(2) = -1$

5. $\dfrac{dy}{dt} = \dfrac{t^2}{y^2}$ $\qquad\qquad$ $y = \sqrt[3]{t^3 + C}$ $\qquad\qquad$ $y(0) = -2$

6. $\dfrac{dy}{dt} = \dfrac{1}{y}$ $\qquad\qquad$ $y = \pm\sqrt{2t + C}$ $\qquad\qquad$ $y(1) = -3$

7. $\dfrac{dy}{dt} = y - y^2$ $\qquad\qquad$ $y = \dfrac{e^t}{e^t + C}$ $\ $ or $\ y = 0$ $\qquad\qquad$ $y(0) = 0.5$

Task 3. Using the solution families and the initial conditions given above in Task 2, find the particular solutions which satisfy the initial conditions, i.e., find C for each one. Do this by hand, first, then check with your CAS.

Task 4. Which of the above DEs have $y(t) = 0$ as a solution, i.e., the trivial solution? Which of them have another <u>constant</u> function as a solution? Did your CAS find these solutions?

- ### Slope fields

    You do not have to solve a differential equation in order to visualize its family of solutions. If you can solve the equation algebraically for dy/dt , you can draw the *slope field*. Learning to do this "by hand" can provide a quick check of what the computer draws and will also give you some insight into the behavior of solutions. So complete the steps in the following example.

Example. Given the DE $t^2 \dfrac{dy}{dt} + y = 0$,

1. Solve the equation for the slope dy/dt.

2. Choose a grid of points (t,y) on the ty-plane (see step 5 below). Often, integer values of t and y are convenient. (Note that in this problem, t can never equal 0.)

3. For each point in the grid, calculate the slope at that point using the DE. For example, since dy/dt = -y/t^2, then at the point (-1,2) the slope is dy/dt = -2/(-1)^2 = -2.

4. Through the point, draw a small line segment having that slope. Some sample line segments are drawn for you in step 5 below. The resulting picture is called the *slope field* or *direction field* for the DE.

5. Finish the slope field on the right below, then sketch in the solution corresponding to the initial condition y(2) = -1 . The solution curves are actually tangent to these small line segments, but can be approximated by letting the small segments run together. What happens to the slopes as t approaches 0 ?

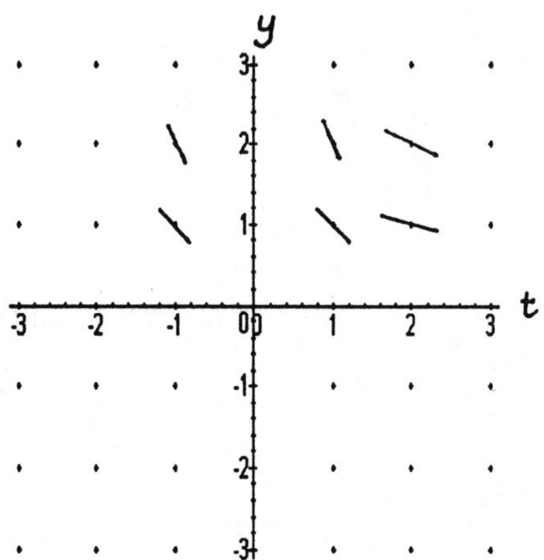

Task 1. For the example you just did above, use your CAS to generate a slope field (see the Appendix for commands).

Task 2. Do two other slope fields from among the problems of Lesson 4, both "by hand" and then with your CAS. Get copies of your fields and sketch the graph of the particular solution which corresponds to the initial condition given in Task 2 of Lesson 4.

- **Euler's method**

From the idea of a slope field as a collection of tangent line segments, we can compute approximate, <u>numerical</u> values for a solution of an initial value problem. That is, we are now seeking *quantitative* information.

The idea is to start a solution at the initial point, then "walk" along its tangent line in order to estimate another point on the solution. This works, since the tangent line stays pretty close to the solution for a short while. We then repeat the process over and over to trace out a solution point by point.

First you'll do a concrete example, then derive the general method.

Example. Start with the initial value problem (IVP):

$$\frac{dy}{dt} = y \ , \quad y(0) = 1$$

Your job is to construct a table of (t,y)-values that will be an approximation of the solution:

t					
y					

1. The first entry in the table is given in the problem statement; find it.

2. What is the slope of the function as it passes through the point in Step 1? (Also found by using the problem statement.)

3. The slope you found in Step 2 will be close to correct for a short time period Δt . In this problem, suppose that $\Delta t = 0.25$. The next t-entry in your table will be: (the old t) $+ \Delta t$. What is it? What is the corresponding new y-entry? (Hint: Use the picture below and the slope found in Step 2, and remember that slope $=$ rise / run.)

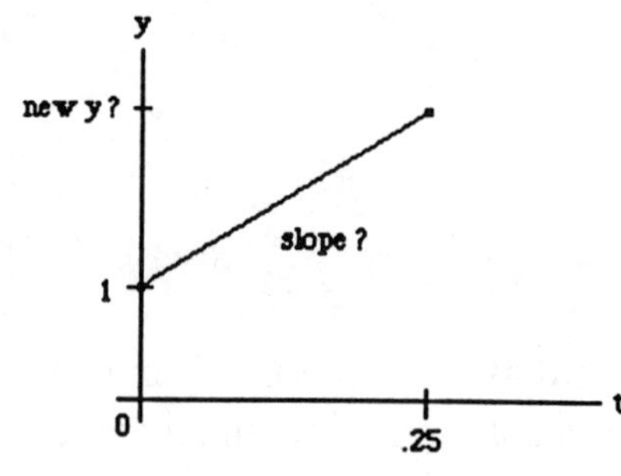

4. To generate the rest of the table, you move from the latest entry to the next using the equation you found in Step 3. Find the next three entries. The t-values will continue in *step-sizes* of $\Delta t = 0.25$, so you'll get $t = 0.50, 0.75, 1.00$. Use a new slope for each t-value to get the next corresponding y-value.

You have just used Euler's Method with step size $\Delta t = 0.25$ to estimate the function e^t on the interval $0 \le t \le 1$, since $y(t) = e^t$ is the solution to the IVP $\dfrac{dy}{dt} = y$, $y(0) = 1$.

Your last y-value in the table is an approximation of $y(1) = e^1 = e$. How close an approximation is it?

Here's the general method. The initial value problem is of the form

$$\frac{dy}{dt} = f(t,y) \ , \quad y(t_0) = y_0 \ .$$

Think of filling in a table such as you did in the example above:

t					
y					

5. What is the first (t,y)-entry in the table, from the problem?

6. What is the slope of the solution as it passes through the point in Step 5 ?

7. As before, the slope in Step 6 will be close to correct for a short time period, Δt. Find the new t-entry, then use the picture below to find the new y-entry.

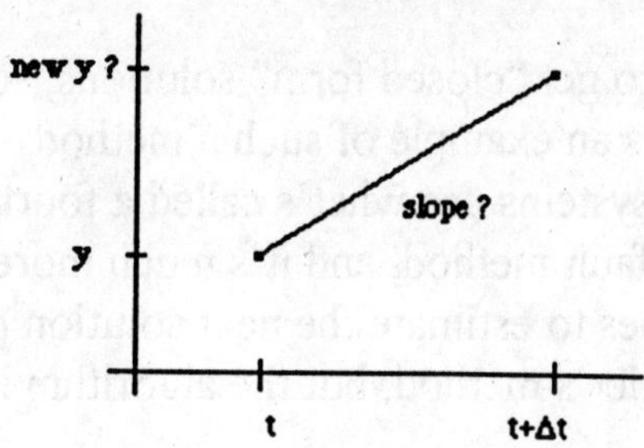

As many entries as you wish can be found by moving one step at a time in this way.

8. Fill in the general formulas:

Euler's Method:

(new t) = ______________;   (new y) = ____________________

The number Δt is called the *step size*.

- **The inaccuracy of Euler's method**

Most DEs cannot be solved to get "closed form" solutions, so you must resort to a numerical method. Euler's method is an example of such a method. Actually, Euler's method is very crude; many computer algebra systems use what's called a fourth-order Runga-Kutta Method (known as RK4) as their default method, and it's much more accurate than Euler. RK4 uses a weighted average of four slopes to estimate the next solution point, and so the geometry of RK4 is hard to see, unlike that of Euler's method, but the algorithm is more accurate than Euler's.

However, since all numerical methods are approximations, we'll stick with Euler for now so that you can more easily see the difference in values between using a numerical approximation and using the "true" solution. Euler's method diverges from the true solution much more rapidly than RK4 does, so the discrepancies between the "true" solution and "numerical" are easier to see.

For completeness, we include here the Euler's method formulas you developed in the last lesson:

$$(\text{new } t) = (\text{old } t) + \Delta t$$
$$(\text{new } y) = (\text{old } y) + f(\text{old } t, \text{ old } y) \cdot \Delta t,$$

where f is the function in the DE $dy/dt = f(t,y)$. The step size, Δt, is usually called h, while $(\text{new } y)$ is called y_n and $(\text{old } y)$ is y_{n-1} (same for the t). So, more mathematically, the Euler formulas for the t-interval $[t_0, b]$ are:

$$t_n = t_{n-1} + h ,$$
$$y_n = y_{n-1} + f(t_{n-1}, y_{n-1})h , \quad \text{for } n = 1, 2, \ldots, k$$

where (t_0, y_0) is the given initial point, and k is the number of steps forward you want to go. The value of k is $(b - t_0)/h$, where h is the step size.

Let's see how accurate Euler's method is. Step through the following example. You'll need to consult the Appendix for appropriate computer commands.

<u>**Example.**</u>   Consider the IVP of Problem 4 from Task 2 of Lesson 4, namely,

$$\frac{dy}{dt} = \frac{t}{y} \quad , \qquad y(2) = -1 \quad ,$$

which has the solution

$$y = -\sqrt{t^2 - 3} \quad \text{for } t \geq \sqrt{3} .$$

1. You're going to make a chart of exact values of the solution vs. the Euler method estimates on the interval $[2, 2.5]$. You'll want to use 5 decimal places in your y-values. Fill in the row marked 'exact y' below. Use your calculator for this, or your CAS.

t	2.0	2.1	2.2	2.3	2.4	2.5
exact y						
Euler y						

2. Now use your CAS to get the Euler approximations for the same t-values, and copy those numbers into their row. Use step size $h = 0.1$.

4. Are the Euler approximations too large or too small? Use the graph of the exact solution to explain why. When graphing, keep the t and y scales the same and use the t-interval $1.9 \leq t \leq 2.6$.

5. You can get more accurate Euler values by decreasing the step size, but only up to a certain point, because a smaller step size means more calculations by the computer, and more calculations mean more round-off error. Use the different step-sizes in the table below to get the Euler values for $y(2.5)$ and fill in the table; use 5 decimal places.

step size h	0.05	0.01	0.005	0.0001
y(2.5)				

6. What is the exact value of $y(2.5)$ to 5 places? Which of the above step sizes came closest?

7. Now plot the exact solution curve and your "best" Euler approximation on the same axes, and compare the results. What do you conclude?

- **A qualitative look at a fish population**

Example. A population of fish is being monitored in a lake. The fish population at any time t we'll call p(t), where p is measured in thousands of kilograms and t is measured in weeks. Then suppose p(t) satisfies the DE

$$(\textit{fishpop}) \qquad\qquad\qquad p' = p - 0.2p^2 - 0.7 \;,$$

where $p' = dp/dt$.

You'll see in a later lesson how such an equation can be determined. For now, you'll do some qualitative analysis of the equation and see what you can predict about the growth or decline of the population using only the DE itself, without knowing the solutions. Fill in the details of the following steps:

1. First, notice that the population doesn't change if $p' = 0$. The values of p at which this happens are called the **equilibrium points**. Find these equilibrium points to 1 decimal place accuracy. (Do this by setting the right-hand side of equation *(fishpop)* equal to zero and then solving the resulting quadratic equation for p .)

    These equilibrium values are **constant solutions** of the DE, that is, they are constant functions of time which are solutions. (Can you verify this?)

2. Now you'll learn how to draw a **phase line** plot of the solutions. A phase line plot is a 1-dimensional plot of a p-axis that shows where the solutions are increasing and decreasing. Use the p-axis shown below, and mark the equilibrium positions on it; you should have found these to be: $p_1 = 0.8$ and $p_2 = 4.2$.

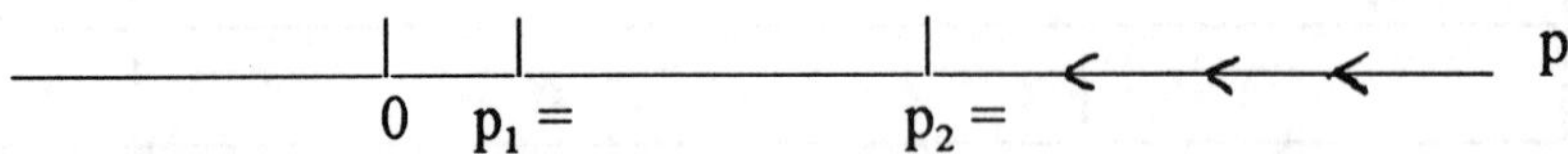

In between the equilibrium values, the population p is increasing if p' is positive, or decreasing if p' is negative. For the values of p for which $p' > 0$, draw small arrows on the phase line pointing to the right, and for $p' < 0$, point the arrows to the left. Some of the arrows have been drawn for you; fill in the rest. (Hint: To figure out where p' is positive or negative, either plug some p values into the right-hand side of *(fishpop)*, or sketch a graph of p' vs. p).

You now have a phase line plot of the solutions, which you can use to get an overview of the behavior. Use it to answer the following questions.

Task 1. Suppose that the lake is initially stocked with 1200 kg of fish. Translate this statement into an initial condition for the DE. Over the next few weeks, what will the population do -- grow? die out? remain the same?

Task 2. Same question as in Task 1, but suppose the initial population is 5000 kg of fish.

Task 3. Draw a phase line plot for the DE

$$\frac{dy}{dt} = 4.1y - y^2 - 4.2 \ .$$

- **A quantitative look at the fish population**

Now you'll study the same fish population as in Lesson 8, but this time you'll answer some <u>quantitative</u> questions.

Recall that the fish population, $p(t)$, satisfies the DE

(*fishpop*) $$p' = p - 0.2p^2 - 0.7 \, ,$$

where p is measured in thousands of kilograms and t is in weeks.

The above DE has two equilibrium populations, $p_1 = 0.8$ and $p_2 = 4.2$, both of which are constant solutions of the equation. Since they're solutions, no other solution can cross through them; this can be seen by applying the Existence and Uniqueness Theorem (see your text) to the problem. Thus, the other solutions either approach them asymptotically as $t \to \infty$, or move away from them. Your phase line plot from the last lesson should show this.

<u>**Task 1.**</u> Use the Euler approximation method with a reasonable step size to answer the following questions.

1. If $p(0) = 6$ (i.e., 6,000 kg), what is the population after 10 weeks?

2. How close is this to the equilibrium point it's approaching?

3. At about what time does the population stop dropping by 100 kg per week?

4. When does the population come within 50 kg of equilibrium?

5. If $p(0) = 0.5$, when does the population die out?

6. If $p(0) = 1.2$, when does the population come within 100 kg of equilibrium?

Task 2. Consider the other DE whose phase line plot you drew in Lesson 8, namely,

$$\frac{dy}{dt} = 4.1y - y^2 - 4.2 \ .$$

Use the Euler method to estimate long-term (t = 10 is probably good enough) behavior for the
following data:

    a) $y(0) = 4$; step size $h = 0.1$
    b) $y(0) = 4$; step size $h = 0.5$
    c) $y(0) = 4$; step size $h = 1.0$

Which results are accurate? Use the phase line plot to explain the accuracy or lack thereof.

Task 3. The differential equation *(fishpop)* can be solved analytically. Do it. Then find the
solution which satisfies $p(0) = 6$ and plot it. What happens to the population as time goes on?

● **Integrating factors for first-order linear equations**

The following examples will show you a new way to look at the product rule for derivatives. This way of thinking will help you to understand how to solve linear DEs in the next lesson. The idea is best explained by example.

Example 1. The product rule for derivatives gives:

$$\frac{d}{dt}(t^2 e^t) = t^2 e^t + 2te^t \ .$$

We like to describe this by saying that

the right side $(t^2 e^t + 2te^t)$ 'folds up' into the left side $\frac{d}{dt}(t^2 e^t)$.

So, if you were to start with another, slightly different, expression, say $(te^t + 2e^t)$, and you asked: "How do I make this expression 'fold up' into the derivative of a product?", the answer would be: "Multiply it by the function t ".

The function t by which you multiply is called an ***integrating factor*** for the expression $(te^t + 2e^t)$. (Don't worry about also having to divide by the function t to keep things equal, since you'll be dealing with equations when you apply this technique.)

Example 2. Assume that y is some unknown function of t , and consider the product $(t^3 y)$. This time we use implicit differentiation and the product rule to get

$$\frac{d}{dt}(t^3 y) = t^3 \frac{dy}{dt} + 3t^2 y \ .$$

Again, you could say that the right side of the above equation 'folds up' into the derivative of $(t^3 y)$.

This means that, if you were to start with the slightly different expression $t\frac{dy}{dt} + 3y$,

and you wanted it to 'fold up' into the derivative of some <u>product</u>, you would have to multiply it by the integrating factor t^2 . The clue to "guessing" this integrating factor is that the "3" in

$t\frac{dy}{dt} + 3y$ is part of the derivative of t^3.

<u>**Task 1**</u>. For each expression given below, find an integrating factor which makes the expression 'fold up' to become the derivative of a product. Assume that y is a function of t.

1. $\quad t^2 \dfrac{dy}{dt} + 3ty$

2. $\quad t \dfrac{dy}{dt} + 5y$

3. $\quad \dfrac{dy}{dt} + y$

4. $\quad \dfrac{dy}{dt} + 2y$

5. $\quad \dfrac{dy}{dt} - 2y$

6. $\quad \dfrac{dy}{dt} + y\cos(t)$

7. $\quad \dfrac{dy}{dt} + p(t)y$

- **Solving first-order linear equations**

A ***first-order linear differential equation*** has the form

$$(1) \qquad a(t)\frac{dy}{dt} + b(t)y = c(t) \, ,$$

where the ***coefficients*** $a(t)$, $b(t)$, and $c(t)$ are usually continuous functions on some t-interval. It is customary to divide through by the leading coefficient, $a(t)$, and write the equation in ***the standard form***

$$(2) \qquad \frac{dy}{dt} + p(t)y = q(t) \, ,$$

where $p(t) = b(t) / a(t)$ and $q(t) = c(t) / a(t)$.

You must know these linear forms and be able to distinguish between linear and nonlinear equations, because the theory of differential equations neatly splits itself into these two camps. Any first-order DE that cannot be put into one of the forms (1) or (2) is ***nonlinear***. The solution method that follows applies only to linear equations.

To solve equation (2), the idea is to find an integrating factor that will make the left-hand side of (2) 'fold up' into the derivative of a product, just as you practiced in Lesson 10. One such integrating factor is

$$(*) \qquad \mu(t) = \exp\left(\int p(t)\,dt\right) \, .$$

Your text will have a derivation of this formula. We'll illustrate how it works with an example. The practice problems you did in Lesson 10 should help you to understand the process.

Example. Solve the DE $\quad t\frac{dy}{dt} - y = 2t^2 + t$.

First, put the equation into standard form by dividing through by t :

$$(3) \qquad \frac{dy}{dt} - \frac{1}{t}y = 2t + 1 \, , \quad for \ \ t \neq 0$$

and identify the coefficients $p(t) = -1/t$ and $q(t) = 2t + 1$. We'll use (*) to find the integrating factor $\mu(t)$. First, integrate $p(t)$ to get

$$\int p(t)\,dt = \int \left(-\frac{1}{t} \right) dt = -\ln|t| \ .$$

Then this gives us our integrating factor

$$\mu(t) = \exp\left(\int p(t)\,dt \right) = e^{-\ln|t|} = e^{\ln|t|^{-1}} = |t|^{-1} \ .$$

If $t > 0$, then $\mu(t) = 1/t$. Multiplying both sides of the DE (3) by $\mu(t)$ gives

$$(4) \qquad \frac{1}{t}\frac{dy}{dt} - \frac{1}{t^2}y = 2 + \frac{1}{t} \ .$$

To see that this integrating factor "works", look closely at the left-hand side of equation (4) and

realize that $\dfrac{d}{dt}\left(\dfrac{1}{t} \right) = -\dfrac{1}{t^2}$, so that the left side 'folds up' into $\dfrac{d}{dt}\left(\dfrac{1}{t}y \right)$. Thus, equation

(4) can be written as

$$\frac{d}{dt}\left(\frac{1}{t}y \right) = 2 + \frac{1}{t} \ .$$

Now integrate both sides of this last equation with respect to t to get

$$\frac{1}{t}y = 2t + \ln t + C \ ,$$

and, finally, solve for y to get the general solution

$$y = 2t^2 + t\ln t + Ct \ , \quad \textit{for} \ \ t > 0 \ .$$

<u>Summary of the method</u>. To solve a linear DE of the form

$$(1) \qquad a(t)\frac{dy}{dt} + b(t)y = c(t) ,$$

a) Divide through by the leading coefficient *a(t)* to put the equation into standard form

$$(2) \qquad \frac{dy}{dt} + p(t)y = q(t) ;$$

b) Determine the coefficient *p(t)* and find an antiderivative of it;

c) Find the integrating factor $\mu(t) = \exp\left(\int p(t)\,dt\right)$;

d) Multiply both sides of the DE (2) above by μ(t) and write the equation in the form

$$\frac{d}{dt}\big(\mu(t)y\big) = \mu(t)q(t) ;$$

e) Integrate both sides (don't forget the constant of integration);

f) Solve algebraically for y .

<u>Task 1</u>. If $t < 0$ in the above example, the integrating factor is $\mu = -1/t$ since, for negative t, an anti-derivative of $1/t$ is $\ln(-t)$. The remaining steps can be completed as above. Do them.

<u>Task 2</u>. Use the method of integrating factors to solve the following linear equations.

1. $\dfrac{dy}{dt} - y = e^{\,t}$

2. $\dfrac{1}{t}\dfrac{dy}{dt} - \dfrac{2y}{t^2} = t\cos t , \quad t > 0$

3. $\left(t^2 + 1\right)\dfrac{dy}{dt} + ty = t$

<u>Task 3</u>. Use your CAS to solve the equations in Task 2. (The necessary solve commands are in Lesson 3 of the Appendix.)

- **Learning to formulate a mixing problem**

Now you'll derive a linear DE that describes a chemical mixing problem. To help you do it, step through the following example. Pay close attention to the <u>units</u> of the quantities involved, because the DE you finally come up with will only be correct if each term in the equation has the same units (it <u>is</u> an equation, after all!)

<u>Example.</u> A tank holds 50 gallons of a salt and water mixture. The tank has intake and outflow pipes which allow flow at the rate of 3 gal/min . We assume that the solution in the tank is of uniform concentration at all times, that is, mixing is instant and thorough. Also assume the tank is full at all times. Under certain conditions on the inflow and outflow, we'll want to know how much salt is in the tank at any given time. Our intent is to formulate and solve a DE that describes the situation.

So let $s(t)$ be the number of <u>pounds</u> of salt in the tank at time t , which is measured in minutes; $s(t)$ is the unknown function we wish to find.

1. Draw a picture of the tank and the intake and outflow pipes. Label the tank with its capacity and the pipes with their flow rates. It always helps to have a picture.

Answer the following questions, using new variables when needed. Be sure to use appropriate units in your answers, like lb/gal or gal/min .

2. What is the <u>concentration</u> of salt in the tank at time t ?
   (Hint: For those of you whose chemistry is rusty, the <u>concentration</u> is the ratio of pounds of salt to gallons of water and is measured in units of lb/gal .)

3. What physical quantity does ds/dt represent? What are its units?

Now suppose the tank is full of salt water. Both pipes are open, and <u>pure</u> water is coming in, while the salt water mixture is leaving.

4. How many <u>pounds of salt</u> are <u>entering</u> the tank each minute? (Don't forget the units.)

5. How many <u>gallons of mixture</u> are <u>leaving</u> each minute?

6. How many <u>pounds of salt</u> are in <u>each gallon</u> of the mixture?   (You've already answered this
    question above, but we're asking it again -- it's the concentration.)

7. How many <u>pounds of salt</u> are <u>leaving</u> the tank each minute?
    (Hint: To get the answer, combine your answers to Step 5 and Step 6 above. Note that
    the units "cancel" just as algebraic symbols do, so that

$$(\textit{concentration}) \times (\textit{flow rate}) = \left(A \frac{lb}{gal} \right) \left(B \frac{gal}{min} \right) = AB \frac{lb}{min} .$$

8. Finally, give a formula for ds/dt . Use your answers to Step 4 and Step 7, and the fact that

$$ds/dt = (\text{net flow rate}) = (\text{flow rate in}) - (\text{flow rate out}).$$

The answer to Step 8 is the desired linear DE in the unknown s(t) , namely,

$$ds/dt = - (3/50)s \ \text{lb/min}.$$

If s(0) were given, you would have an IVP which you could solve. In other words, if you knew
how much salt was initially in the tank, you could now figure out how much salt there is at <u>any</u>
time t . Let's do it.

9. Suppose the <u>initial concentration</u> of salt in the tank is 0.5 lb/gal . What is s(0) ?

10. The initial value problem you should have from Steps 8 and 9 is   ds/dt = - (3/50)s lb/min
     with s(0) = 25 lb. Solve this IVP and thus find s(t) , the amount of salt at any time t .

Now that you know the amount of salt at any time t , you can answer questions like the
following.

11. When will the concentration be cut in half?

12. When will the water in the tank be completely free of salt?

●　　**A harder mixing problem**

In Lesson 12, you formulated a DE for $s(t)$, the number of pounds of salt in a tank at time t minutes. You did this by using "common sense" to realize that:

the net rate of change of s = (input rate of s) - (output rate of s),

or, more simply,

(*Principle* 1)　　　　　　$$\frac{ds}{dt} = input\ rate - output\ rate\ .$$

All three terms of this DE, namely, *ds/dt*, *input rate*, and *output rate*, have to be measured in the same units like lb/min or kg/sec. For mixing problems, the *input rate* and the *output rate* are products of concentrations and flow rates, as in Question 7 of Lesson 12.

You'll see that *(Principle 1)* is a useful general principle to use in formulating DEs for several different kinds of applications: heat flow and population problems are two examples.

Work on the following mixture problem, which is a little more involved than the previous one. The steps outlined in the questions following the problem statement will help you derive the correct IVP.

Example. A 50% alcohol solution flows into a tank at the rate of 6 liters/min. The tank initially held 100 liters of a 10% alcohol solution. The well-stirred solution flows out at the rate of 8 liters/min. Find the amount of alcohol in the tank at any time t.

Answer these questions to guide yourself to the solution:

1. Draw a picture of the tank and the intake and outflow pipes. Label the tank with its capacity and the pipes with their flow rates.

2. Set up some notation: Let $x(t)$ be the (unknown) number of liters of alcohol in the tank at time t minutes

3. You'll soon need to know the concentration of alcohol at any time t, and for that you'll need the volume of fluid at any time t. So let $V(t)$ be the (unknown) volume of fluid in the tank at time t; V changes with time, since the inflow and outflow rates are different. Write a DE involving V and dV/dt using Principle 1 above, with V instead of s. Remember, you are measuring only the amount of the <u>combined</u> fluids.

4. Since $V(0) = 100$ liters , you can solve the DE of Step 3 for $V(t)$. Do so. (You should get
    $V(t) = 100 - 2t$ liters.)

5. Find the concentration of alcohol in the tank at time t .

6. What is the input rate for alcohol?

7. What is the output rate for alcohol?

8. Finally, use Step 6 and Step 7 and Principle 1 to write the DE for the unknown $x(t)$. What
    is the initial condition on x , i.e., what is $x(0)$?

   The IVP you should end up with is

$$\frac{dx}{dt} = 6 \times (0.5) - \frac{8x}{100 - 2t} \; \frac{liters}{min}, \quad x(0) = 10 \; liters .$$

If you didn't get this answer, go back and figure out where you went wrong. ☹

Task 1. Solve the above IVP by hand; it's linear. Then use your CAS to solve it.

Task 2. Use your CAS to graph the solution of the IVP. At what time does the solution fail to
describe the actual physical problem? Modify your graph (by hand) to show what really happens.

Task 3. Graph the concentration function, $x(t)/V(t)$.

Task 4. What is the limit of the concentration as $t \to 50$?

Task 5. Does the percentage of alcohol in the tank ever reach 40% ? Explain. What is the
maximum percentage of alcohol reached?

● **Formulating a population problem**

Now you're going to apply Principle 1 from the last lesson to a population problem. We'll remind you that the principle is:

(the net rate of change of p) = (input rate of p) - (output rate of p) ,

where this time $p = p(t)$ is the population size at time t , the net rate of change of p is dp/dt , the input rate is the number of births per unit time, and the output rate is the number of deaths per unit time.

Doing the tasks that follow should give you some insight into how to think about population problems.

A population of size p changes with time t . We will leave the units unspecified for now. They could be in individuals, millions of individuals, or grams. Time t could be in years, weeks, or minutes. But we assume that p is a <u>continuous</u> function of time. (This necessitates an approximation when p is integer-valued.)

Task 1. Write the DE which reflects each of the following sets of assumptions:

1. The number of births per time unit (i.e., the birth rate) is constant; no deaths.

2. The number of births per time unit is proportional to population size; no deaths.

3. Both birth and death rates are proportional to population size.

Task 2. Draw approximate graphs of p vs. t which correspond to each of the assumptions in Task 1.

Task 3. Using the assumptions of Task 1.3 , draw an approximate graph of the following:

1. the birth rate vs. population size;

2. the death rate vs. population size.

Task 4. Now assume the death rate is <u>not</u> proportional to population size.

1. Draw a graph of the death rate vs. population size which represents an escalating death rate
    (perhaps due to overcrowding).

2. Make up a specific algebraic rule for the graph in part a; that is, if d is the death rate,
    write a function d(p) whose graph looks like the function of part 1.

Task 5. In Lessons 8 and 9 you studied a fish population modeled by the DE

   (*fishpop*) $p' = p - 0.2p^2 - 0.7$.

Use what you learned in Tasks 1 - 4 above to explain each term on the right-hand side of the DE
(fishpop) in terms of birth and death rates.

- **The Malthusian population growth model using laboratory data**

If we assume that a population has ample food and space in which to grow, then we can assume that it grows at a rate proportional to the population that is present, and that the death rate is zero, at least in the short run.

This idea leads to the **Malthusian** or **exponential model** for population growth:

$$\frac{dp}{dt} = rp \,, \quad p(0) = p_o \,,$$

where, as usual, $p(t)$ is a measure of the size of the population at time t, r is the proportionality parameter for the growth rate (it's a constant which is characteristic of the particular population) and p_0 is the initial population.

In this lesson you'll construct and solve a DE model for a population of bacteria. Do the following steps.

1. Find the general solution of the Malthusian model given above. You've solved it before, as a separable equation. This time, solve it as a linear equation.

Below is the data from a biology experiment in which bacteria were being cultured in a flask. The population was actually measured with a spectrometer, so the units are those of light refraction rather than population size. But the two are proportional, so you may treat the numbers as a measure of population size. Time is in hours.

t	0.0	0.5	1.0	2.0	3.0	4.0	6.0
p(t)	0.020	0.024	0.031	0.043	0.062	0.080	0.163

2. Use your CAS to plot a graph of the above points.

3. The shape of the graph should suggest a Malthusian growth model. Does it? Write the DE and initial condition for this model. The parameter r is, as yet, undetermined, so leave it as just r.

4. Solve the IVP of Step 3. You may wish to use your result from Step 1.

5. Your solution in Step 4 will still have r in it. Use more information from the data to find a value for it.

6. You should now have a complete solution, $p(t)$. Plot it and the data points on the same axes. How good a fit do you have?

7. Remember that real data can be messy; your function may not fit perfectly. Adjust the parameters in your solution to see if you can get a better fit. Write the function you found which best approximates the data. Get a plot of it and the data points.

8. What does your model predict about the population size as time goes on? If the culture were allowed to grow for a much longer period of time, what adjustments might you make in the model?

- **The logistic population growth model using laboratory data**

Recall the bacteria population of Lesson 15. If the bacteria is allowed to grow for a total of 12 hours, you will want to use a different model, because overcrowding will start to occur.

To see how "overcrowding" can be incorporated into a DE, we ask the question: "In how many ways can creatures (humans or bunnies or bacteria) meet each other?" This can happen in $p(p-1)/2$ ways. (For example, in a group of 10 people, each one can shake hands with the other 9 in $10{\cdot}9$ or 90 handshakes, but this counts Jim shaking hands with Jane and Jane shaking hands with Jim as separate handshakes, so we must divide by 2 .) So we estimate that the death rate is proportional to $p^2 - p$. This observation, combined with a growth rate proportional to p leads to *the logistic model* for population growth:

$$(1) \qquad \frac{dp}{dt} = ap - bp^2 , \quad p(0) = p_o ,$$

where a and b are population parameters (constants) to be determined from the particular population. To study this model, answer the following questions.

1. Equation (1) is separable and nonlinear. Find its general solution.

You should have obtained the solution

$$p(t) = \frac{ap_o}{bp_o + (a - bp_o)e^{-at}} ,$$

which is called a *logistic function.*

A typical logistic curve is shown below. Its shape is a so-called "sigmoidal" or "s-shaped" curve.

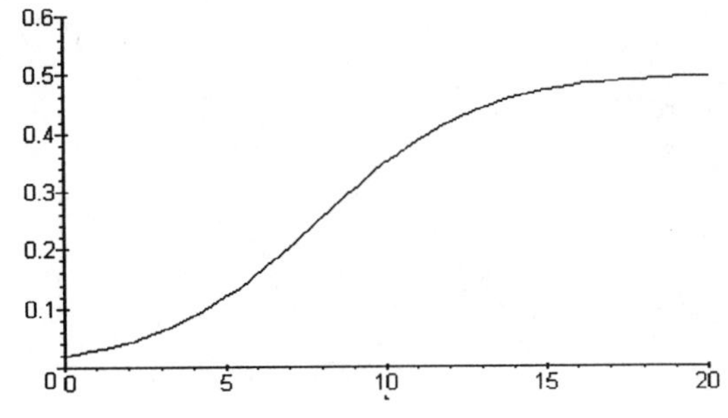

2. What are the equilibrium populations for the logistic equation (1)? What happens to the
   population as t gets large?

    The limiting value of $p(t)$, namely a/b , is referred to as the ***carrying capacity***. It's the
maximum population the environment will allow.

    The values of a and b can be shown to be:

$$(2) \qquad a = \frac{1}{t_1} \ln\left[\frac{p_2(p_1 - p_0)}{p_0(p_2 - p_1)}\right] , \qquad b = \frac{a}{p_1}\left[\frac{p_1^{\,2} - p_0 p_2}{p_1 p_2 - 2p_0 p_2 + p_0 p_1}\right] ,$$

where $p_1 = p(t_1)$ *and* $p_2 = p(t_2)$ *are given data points with* $t_2 = 2t_1$ *and* $t_1 > 0$.

Now you'll fit a logistic curve to the data of Lesson 15.

3. Which pair of data points from Lesson 15 do you think will give the best values for the
   parameters a and b ? Use these points and the formulas (2) above to find a and b to 3
   decimal places, and then use a and b to find $p(t)$ for this data.

4. Plot $p(t)$ and the data points on the same axes. How good a match is it? It should be fairly
   good. If not, go back and do it again using different data points.

5. What does your solution predict for the population size at t = 12 hours? What happens to
   $p(t)$ as t gets large? Does this number agree with the predicted value of a/b ?

- **Newton's second law and falling bodies**

Isaac Newton was not only one of the inventors of calculus, but he was the first to write down and study a differential equation. In investigating the motions of the planets, he formulated his three famous laws. One of these, his Second Law, is of particular importance in the study of differential equations.

Newton's Second Law:

$$m \cdot a = \sum F_i \, ,$$

where m = mass of an object, a = acceleration of the object, and F_i is a force acting on the object. Acceleration and forces are <u>vectors</u>, so $\sum F_i$ is the <u>vector sum</u> of all the forces acting on the object.

You're going to study the motion of objects falling near the surface of the Earth. When an object is near the surface of the Earth, it is acted upon by gravity in such a way that its <u>acceleration</u>, if no other force is present, is <u>constant.</u> Thus we have the following important numbers representing the acceleration due to gravity near the Earth's surface:

$$g = 32.16 \text{ ft/sec}^2 = 9.80 \text{ meters/sec}^2 \qquad \text{(in Ithaca, NY)}$$

The value of g depends on the units of measurement you choose to use. In the British system, acceleration is measured in ft/sec^2, while in the S.I. (System Internationale) system, acceleration is measured in meters/sec^2.

Step through the following questions to gain some understanding about how to set up a differential equation for a falling object.

1. In the United States, most people (but not scientists) measure the <u>weight</u> of an object in units of pounds, and this represents the <u>force</u> of gravity acting on the object. Use Newton's Second Law to find the <u>mass</u> of an object which weighs 160 pounds. (The unit of mass in the British system is the picturesque "slug".)

2. In the rest of the world, the mass of an object is measured in kilograms. Use the Second Law to find the <u>force</u> due to gravity acting on an object of mass 50 kg. (The unit of force in the S.I. system is the "newton", named for Sir Isaac.)

3. Assume an object is moving in a straight line, say on the x-axis. If its position at time t is
   x(t) , what calculus symbols represent the velocity and acceleration of the object?

   Here's a problem you may recognize from calculus:

4. An object is released from rest from a height of 152 feet above the Earth's surface and
   allowed to fall straight down under the influence of gravity. Assume no other force acts.
   Draw and label a picture describing its position as a function of time, and use Newton's
   Second Law to write an initial value problem describing this situation.

   For convenience, set up your coordinate axis as follows: Let the origin be at the initial
position of the object, and take the positive direction of the position axis to be <u>downward</u>, in the
direction of motion. Typically, s is the name of the function which describes the position of the
object as a function of time, so use this name. Remember that this is one-dimensional motion:
just down or up, not sideways.

> <u>Warning</u>: It is important that you label your axis appropriately, and stick
> with that choice throughout the problem. Otherwise, your formulas will be
> incorrect.

5. What is the order of the DE in Step 4? How many initial conditions are there? There should
   be two. Write them down.

   With the axis set up as described above, you should have derived the following second-
order DE with two initial conditions:

$$\frac{d^2s}{dt^2} = 32 \ \frac{ft}{sec^2} \ , \quad s(0) = 0 \ ft \ , \quad v(0) = 0 \ \frac{ft}{sec} \ .$$

Can you explain how the initial conditions were obtained?

6. Solve the above initial value problem.

7. When does the object hit the ground? With what velocity does it hit?

   Now let's use the greater power of differential equations to make a more sophisticated
model.

8. If you actually dropped an object from a height of 152 feet, it would take longer to hit the
    ground than the 3.07 seconds predicted by the previous model. Why?

9. One way to describe the force of air resistance is to assume that it is more influential at
    faster speeds than at slower speeds. If F_a represents the force of air resistance, and v is
    the velocity, write an equation that says that air resistance is proportional to velocity.

10. Now write a DE for this same falling object in terms of the velocity function v = v(t) instead
    of s(t) . This equation will be first order instead of second order. To guide you in your
    thinking, do the following:

    a. Draw and label another picture, again taking the positive direction to be downward and
    choosing the origin to be at the initial position of the object.

    b. Write Newton's Second Law in terms of v , assuming gravity and air resistance are the
    forces acting. Remember, the forces are vectors, and air resistance will be in the direction
    opposite to the force of gravity.

11. How many initial conditions go with this DE? Write it or them.

    The above steps should have brought you to the following IVP:

$$m\frac{dv}{dt} = mg - kv , \quad s(0) = 0 , \quad v(0) = 0 .$$

The constant of proportionality, k , in the DE is called the *coefficient of air resistance*, and
differs from object to object. It can be determined experimentally.

12. What kinds of objects do you think have low coefficients of air resistance, and what objects
    have higher ones? What features of an object determine its coefficient of air resistance?

13. Solve the IVP of Step 10b for the function v , including the constant. Integrate v to get the
    position function s and find its constant as well.

14. The above IVP consists of a DE and two initial conditions. What units would be appropriate
    for each of the three equations in the British system? In the S.I. system?

- **Warm-up exercises for second order linear equations**

This lesson will give you practice with typical solutions of some second order DEs.

Task 1. For each function below, write its first and second derivative. Do these by hand and then check with your CAS if you need to.

$f(t)$	$f'(t)$	$f''(t)$
1. e^{t}		
2. e^{-t}		
3. $3e^{-t}$		
4. e^{-3t}		
5. $\sin(t)$		
6. $\cos(t)$		
7. $4\sin(t) - 5\cos(t)$		
8. $\sin(2t)$		
9. $7\sin(5t)$		
10. $e^{-t}\cos(3t)$		

Task 2. Now figure out which of the functions above are solutions of the second order linear homogeneous DEs given below. On what t-intervals are these solutions good?

$$\underline{\hspace{3cm}} \quad \frac{d^2y}{dt^2} + y = 0 \qquad\qquad \underline{\hspace{3cm}} \quad \frac{d^2y}{dt^2} - 9y = 0$$

$$\underline{\hspace{3cm}} \quad \frac{d^2y}{dt^2} + 25y = 0 \qquad\qquad \underline{\hspace{3cm}} \quad \frac{d^2y}{dt^2} + ky = 0$$

$$\underline{\hspace{3cm}} \quad \frac{d^2y}{dt^2} + 4y = 0 \qquad\qquad \underline{\hspace{3cm}} \quad \frac{d^2y}{dt^2} + k_1\frac{dy}{dt} + k_2y = 0$$

$$\underline{\hspace{3cm}} \quad \frac{d^2y}{dt^2} - y = 0$$

Task 3. Do you notice any patterns in the above equations and their solutions? If so, describe the patterns; if not, look again.

Task 4. Find as many solutions as you can for each of the following DEs.

1. $\dfrac{d^2y}{dt^2} - 4y = 0$

2. $\dfrac{d^2y}{dt^2} + 36y = 0$

● A first look at a simple mass-spring system

Scientists and engineers need to study vibrations in many different kinds of mechanical systems; for example, the motion of an automobile body as it moves along a bumpy road, or the motion of an airplane's wings in flight. To begin your study of vibrating systems, you'll look at a simple mechanical system consisting of a mass on the end of a coil spring.

Imagine a spring suspended from a sturdy beam, as in the figure at right, and a weight hung on the spring. You pull the weight down (or push it up) vertically, then let go. The weight will move up and down in a vertical line. You're going to study the motion of this weight.

Set up a coordinate system as follows: Place a y-axis along the direction of motion of the weight, with its positive direction <u>upward</u>. Make the origin $y = 0$ on the y-axis correspond to the "equilibrium position", that is, where the weight would be when it is not moving (the system is "at rest").

Let $y(t)$ designate the <u>position</u> of the weight at time t . This is the unknown function we wish to find.

1. Imagine pulling the weight down and then letting it go. Draw an approximate graph of what you expect the position y of the weight as a function of time t to look like.

2. Use your CAS to graph each of the ten functions of Task 1 in Lesson 18 on at least the interval $0 \le t \le 4\pi$. Pretend that each function is the graph of the position function of a weight suspended on a spring, and describe in physical terms the motion of the weight. Tell what the position and velocity of the weight are at $t = 0$, i.e., the initial position and velocity. Use the notation y_0 and v_0 for these.

3. Do you think your descriptions in Step 2 are realistic, that is, could mass-spring systems really behave like these functions?

- **Deriving the differential equation for the mass-spring system**

Now you'll derive the differential equation which is most often used to model the motion of a mass on the end of a spring. As before, the position of the mass at time t is given by the function $y(t)$. Take the positive y-axis <u>upward</u> with equilibrium position at $y = 0$.

Follow the steps below to guide you in the derivation of the DE.

1. What calculus symbols represent the velocity and acceleration of the mass?

2. You'll need Newton's Second Law. Write it down (see Lesson 17).

Let's look at the forces acting on the mass. The primary force is the <u>restoring force</u>, F_R, of the spring. When the object is above the equilibrium position, the spring is compressed and attempts to push the object down toward equilibrium. Similarly, when the object is below the equilibrium position, the spring is stretched, and it pulls the object upward, again toward equilibrium. The greater the compression or stretching, the larger this force is. A simple model for this force was written by Robert Hooke (1635-1703) and is given by:

<u>Hooke's Law</u>: The restoring force is proportional to
the displacement from equilibrium.

3. Since the displacement of the mass at time t is just y, use Hooke's Law to write an expression for the restoring force, F_R. Remember that this force is a vector and will have direction. The unknown constant of proportionality is called the <u>spring constant</u> (every spring has its own), and is usually denoted by k and taken to be positive.

Gravity, of course, is another force acting on this object, but we'll assume it is exactly counterbalanced by the spring tension. This means that the mass will hang in the air in its equilibrium position indefinitely, without falling to the ground. For this reason you will see no gravity term in your DE. (This is an oversimplified argument; for the true story, see a physicist.)

4. A third force is that due to <u>air resistance</u>, F_A. Assume the force of air resistance is proportional to the velocity of the object. This force is also a vector and acts in the direction which opposes the motion of the mass. So, if y' is positive, F_A must be negative, and vice

versa. Write an expression for this force. Denote the constant of proportionality here by b. (It is also known as the <u>coefficient of damping</u>.) Make sure that your force reflects the fact that $b > 0$.

5. Thus, there are two forces acting on the object, F_R and F_A. Let the object's mass be m (we'll ignore the mass of the spring). Now use Newton's Second Law to write the differential equation which models the motion of the mass on the end of the spring.

6. What order is this DE?

7. How many initial conditions will be important? What are they?

You should have arrived at the DE

$$ma = -F_A - F_R$$
$$= -bv - ky$$

or, equivalently, since $v = dy/dt$ and $a = d^2y/dt^2$,

$$(spring) \qquad m\frac{d^2y}{dt^2} + b\frac{dy}{dt} + ky = 0 \, ,$$

and you need to know the initial values (conditions) $y(0) = y_0$ and $y'(0) = v_0$ in order to describe a solution exactly.

So you see that the exact position function, $y(t)$, of an object vibrating on a spring depends upon a number of things:

m = the mass of the object
b = the coefficient of air resistance ("damping")
k = the stiffness of the spring (the "spring constant")
y_0 = the initial position of the object
v_0 = the initial velocity of the object

Equation *(spring)* is another example of a ***second-order linear homogeneous differential equation with constant coefficients.*** You saw other examples of this form of equation in Lesson 18. The word "homogeneous" refers to the fact that ***the zero function,*** $y(t) = 0$, is a solution of the DE, the so-called ***trivial solution.*** (The significance of the word "linear" will become clear later.)

This form of equation is one of the few second-order DEs that we're able to solve in closed form, and you'll learn how to do that soon.

- **Entering the 'spring' program into your CAS**

In Lesson 20 you developed the second-order linear homogeneous constant coefficient DE which is most often used to model the motion of a mass on the end of a spring. That equation is:

$$(spring) \qquad m\frac{d^2y}{dt^2} + b\frac{dy}{dt} + ky = 0 \; ,$$

where m is the mass of the object, b is the coefficient of air resistance, and k is the spring constant. All of these parameters are positive. The positive y-direction is upward.

To completely determine a unique solution of equation *(spring)*, you need to specify initial conditions y_0 and v_0, that is, the initial position and velocity of the mass.

Now you'll get ready to use your CAS to investigate the behavior of solutions of *(spring)* for various values of the parameters m, b, k, y_0, and v_0.

Task 1. The Appendix contains a little program named 'spring' which will draw graphs of the position function $y(t)$ of the mass for various values of the parameters that you choose to use. You'll only need to type in the entire program the first time you use it; then save it to a diskette for future use. So type in the procedure, <u>exactly</u> as it is written in the Appendix, and save it on your diskette.

Task 2. To test your program, type and execute the command

$$spring(1, 0, 1, 1, 0) \; .$$

You should get a graph of the cosine function on the time interval $[0,15]$. If you didn't get this, either the program was entered incorrectly, or the command was. Check your work.

Task 3. Program 'spring' is written so that it accepts as input the 5 parameter values:

> m = the mass of the object,
> b = the coefficient of air resistance ("damping")
> k = the stiffness of the spring (the spring constant)
> y_0 = the initial position of the mass,
> v_0 = the initial velocity of the mass,

<u>in that order</u>. For an object of mass 1 , no air resistance, spring of stiffness 2 , initial position 3 units <u>below</u> equilibrium, and initial velocity 4 (in the up direction), you would type

$$\text{spring}(1, 0, 2, -3, 4) \, ,$$

Do so. Does this picture agree with the behavior you would intuitively expect?

• **Investigating the behavior of a mass-spring system: No damping**

Now you're ready to do some experimenting with different masses and stiffnesses. First, you'll change just the mass, m, while keeping the other parameters the same. Then you'll change just the spring stiffness, k, and observe what happens. Throughout these experiments, the damping factor, b, will be kept at 0 so as to keep things simple; also, y_0 will be kept at 1 and v_0 will be kept at 0.

The <u>basic system</u> will be

$$\text{spring}(1, 0, 1, 1, 0)$$

which gives a graph of the function $y = \cos(t)$ (see Task 2 of Lesson 21). The cosine function is the unique solution of the initial value problem $y'' + y = 0$, $y(0) = y_0 = 1$, $y'(0) = v_0 = 0$. Here, $m = 1$, $b = 0$, and $k = 1$, which is why the computer command is $\text{spring}(1,0,1,1,0)$.

<u>Note to the Instructor</u>. If your class is doing this Lesson in groups, we have found it helpful to divide the class into two groups: the mass group and the spring stiffness group, M and K for short. The M group can do Tasks 1, 2, and 3, and gather information on how changing the mass affects the behavior, while the K group can do Tasks 1, 4, and 5 on changing the stiffness. Then each group can report their results to the other group.

Task 1. Type the command for the basic system again, so you are familiar with its graph.

Task 2. To simplify your work, define a function of one variable, the mass m, so that you can quickly enter different masses into the basic system while keeping the other parameters the same. See the Appendix for the syntax.

Task 3. Now enter different masses and compare the results to the basic system, as follows:

    a. Experiment a bit until you find how much you should change the mass, then <u>decide on a systematic way to record the results</u> of various masses, in some helpful order.

    b. What changes in each picture, and what stays the same? In particular, how does the period of the motion change? the amplitude?

    c. Is there a change in the general behavior of the object, or just in the details?

    d. Can you guess at a formula for predicting the changes you see? (This formula is simple but not obvious. We'll be able to prove it later.)

Now you'll repeat the above experiment, this time varying the stiffness k.

Task 4. Define a function of the one variable k so that you can quickly enter different stiffnesses. See the Appendix for the syntax.

Task 5. Now enter different stiffnesses and compare the results to the basic system. Repeat parts a. through d. of Task 3.

You should now be able to answer the following questions:

1. What is the effect of quadrupling the mass? quadrupling the stiffness? quartering the mass? quartering the stiffness?

2. Describe the graph when $m = 10$ and the other parameters are the same as in the basic system.

3. Find a k-value whose graph exacly matches the graph in Question 2, while the other parameters are the same as in the basic system.

4. Let T be the period of the mass' motion, i.e., T is the time it takes for the mass to return to its original position and velocity. Below are some pairs of period T vs. mass m for solutions of the spring equation with the other parameters as in the basic system. There is a function relating T and m, call it $T = f(m)$. See if you can find f.

m	1/16	1/9	1/4	1	4	9	16
T	$\pi/2$	$2\pi/3$	π	2π	4π	6π	8π

5. Below are some pairs of period T vs. stiffness k for solutions of the spring equation with the other parameters as in the basic system. There is a function relating T and k, call it $T = h(k)$. See if you can find h.

k	16	9	4	1	1/4	1/9	1/16
T	$\pi/2$	$2\pi/3$	π	2π	4π	6π	8π

- **Solving a second order linear homogeneous equation: Real roots**

In this lesson you'll learn how to solve some second-order equations to get "closed-form" solutions.

The spring equation

$$(1) \qquad\qquad my'' + by' + ky = 0$$

is a typical **second order, linear, homogeneous, constant coefficient differential equation.**

Look at equation (1) which says that a solution function $y(t)$ has the property that a linear combination of it and its first two derivatives is zero for all t on some interval. (The expression $my'' + by' + ky$ is called a **linear combination of y, y', and y''.**) You'll use this observation to investigate the kinds of solutions equation (1) might have.

1. Does $y = t^2$ have the property that a linear combination of y, y', and y'' adds up to the zero function? Try it.

2. Does $y = 1/t$ have this property? Try it.

3. What is the most reasonable function which has this property? (Hint: What function do you know that looks similar to its first two derivatives?)

4. Let $y = e^{rt}$, where r is a constant. Find its first two derivatives with respect to t and plug them all into equation (1).

5. Simplify and solve the equation you get, remembering that the exponential function is never zero. Which values of r "work" ?

When you substituted $y = e^{rt}$ into the DE (1) in Step 4, you should have ended up with

$$e^{rt}(mr^2 + br + k) = 0 \, ,$$

and, since e^{rt} is never zero, you must therefore conclude that

$$(2) \qquad\qquad mr^2 + br + k = 0 \, .$$

Equation (2) is called the **auxiliary equation** or **characteristic equation** of the DE (1), and it's just a second-degree polynomial in the variable r. Using the quadratic formula to solve (2) gives the two solutions (called **eigenvalues** or **characteristic roots**)

$$r_1, r_2 = \frac{-b \pm \sqrt{b^2 - 4mk}}{2m}.$$

(This is the answer to Step 5 above.) If these eigenvalues are distinct ($r_1 \neq r_2$), then they'll give two different solutions of the DE (1), namely,

$$y_1(t) = e^{r_1 t} \qquad and \qquad y_2(t) = e^{r_2 t}.$$

Task 1. Find two different solutions for each of the following DEs by using the above method to find the eigenvalues. Then plug each function you get into its DE to verify that it's a solution.

1. $\quad y'' - 6y' + 8y = 0$
2. $\quad y'' + 2y' - y = 0$

What happens if the eigenvalues are <u>equal</u>? (This will happen when the **discriminant** $b^2 - 4mk$ is zero.) Then the method produces only <u>one</u> solution, $y_1(t) = e^{rt}$. However, a second, different solution will always be $y_2(t) = t e^{rt}$.

Task 2. For each of the following DEs,
    i. Find the characteristic equation and the <u>repeated</u> eigenvalues, r ;
    ii. By plugging it into the DE, show that $y_2(t) = t e^{rt}$ is also a solution.

1. $\quad y'' - 4y' + 4y = 0$
2. $\quad 4y'' + 20y' + 25y = 0$

The case when the discriminant $b^2 - 4mk$ is negative will give two <u>complex</u> roots. You'll learn how to handle these soon.

- **Working with complex numbers**

As you pursue one mathematical field (in this case, we are studying <u>real</u>-valued differentiable functions of <u>real</u> variables), it is an inevitable and interesting fact that you are drawn into other fields. A little knowledge of <u>complex</u> numbers will help us tremendously in solving differential equations.

You saw in the last lesson that the spring equation will have complex characteristic roots when the discriminant $b^2 - 4mk$ is negative. Do the following Tasks to get ready to solve this case of the spring equation.

Remember that a ***complex number*** is a number of the form

$$a + bi$$

where a and b are real numbers and $i^2 = -1$. To perform a complex arithmetic operation means to put the answer into this ***standard form*** $a + bi$.

Task 1. Evaluate the following and put your answers in standard form:

1. $(3 - 7i) + (8 + 2i)$
2. $(-1 - i) - (3 + 5i)$
3. $4i(3 + 9i)$
4. $(8 + 2i)(7 + i)$ (*FOIL*)
5. $\dfrac{6}{2 + 6i}$ (*Use the conjugate* $2 - 6i$)

Task 2. Solve the following quadratic equations for z. Put your answers in standard form.

1. $z^2 + 9 = 0$ 3. $z^2 + 2z + 3 = 0$

2. $z^2 + 8z + 6 = 0$ 4. $3z^2 - 2z + 4 = 0$

Task 3. Simplify the following:

1. $\cos(-b) + i{\cdot}\sin(-b)$ (*Hint: cosine is even and sine is odd.*)

2. $\dfrac{1}{2}[\cos(b) + i{\cdot}\sin(b)] + \dfrac{1}{2}[\cos(b) - i{\cdot}\sin(b)]$

3. $\dfrac{1}{2}i[\cos(b) - i{\cdot}\sin(b)] - \dfrac{1}{2}i[\cos(b) + i{\cdot}\sin(b)]$

From Taylor Series, it can be shown that

$$e^{ib} = \cos(b) + i{\cdot}\sin(b) ,$$

which is a complex-valued function. This equation is known as ***Euler's formula***, and is our link to differential equations.

Task 4. Use Euler's formula to write the expressions in Task 3 in terms of the complex exponential function, e^{z}.

- **Solving a second order linear homogeneous equation: Complex roots**

To understand how to get <u>real</u>-valued solutions when a DE has <u>complex</u> eigenvalues, work through the steps of the following example.

Example. Consider the DE $y'' + 4y = 0$.

1. Find the characteristic equation and the roots of this DE (see Lesson 23).

2. Write two different solutions of the above DE as complex exponential functions of t (they are not real-valued functions.) Call your solutions z_1 and z_2 .

3. Use Euler's formula (see Lesson 24) to write the functions you found in Step 2 in terms of the sine and cosine. (Use the fact that cosine is an even function and sine is odd.)

4. Use the results from Step 3 to find and simplify the following two functions:

$$y_1 = (\tfrac{1}{2})\, z_1 + (\tfrac{1}{2})\, z_2 \quad \text{and} \quad y_2 = (\tfrac{1}{2})\, i\, z_2 - (\tfrac{1}{2})\, i\, z_1 \; .$$

5. The functions y_1 and y_2 from Step 4 should be the <u>real</u>-valued functions $y_1 = \cos(2t)$ and $y_2 = \sin(2t)$. Plug each of these into the DE and verify that each is a solution.

Step 4 of the above example shows how the two <u>complex</u>-valued solutions $z_1 = e^{2it}$ and $z_2 = e^{-2it}$ can be combined, using linear combinations of them, to produce two different <u>real</u>-valued solutions. It's rather magical!

In the above example, the eigenvalues were the ***pure imaginary*** numbers $\pm 2i$. What do you do if they're not pure imaginary? The following Task will answer the question.

Task 1. Use laws of exponents and Euler's formula to show that

$$e^{(a+bi)t} = e^{at}\cos(bt) + i \cdot e^{at}\sin(bt) \; .$$

Task 2. Find two different real-valued solutions y_1 and y_2 of the DE $y'' - 6y' + 13y = 0$ by repeating Steps 1 - 5 of the example. In Step 3, make use of the result in Task 1. (You should end up with $y_1 = e^{3t}\cos(2t)$ and $y_2 = e^{3t}\sin(2t)$.)

Task 3. Find two different real-valued solutions of the DE $y'' - 6y' + 10y = 0$. Then plug each one into the DE to verify that it really is a solution. You will find that your CAS is a real time-saver here for the algebra.

•　Linear independence and the general solution of the linear homogeneous equation

You might have noticed in Lessons 23 and 25 that a second order linear homogeneous constant coefficient DE has what we've been calling two "different" solutions. The word "different" means that neither solution is a constant multiple of the other one, so that each solution contains different information.

For example, the DE $y'' + y = 0$ has the two "different" solutions $y_1 = \cos(t)$ and $y_2 = \sin(t)$, and neither function is a constant times the other. On the other hand, the function $y_3 = 17\sin(t)$ is also a solution of the equation, and you can see that $y_3(t) = 17 \cdot y_2(t)$, that is, y_3 is a constant multiple of y_2 .

Any two solutions, $y_1(t)$ and $y_2(t)$, of a second order linear homogeneous DE are called *linearly independent* if neither is a constant multiple of the other. So $y_1 = \cos(t)$ and $y_2 = \sin(t)$ are a linearly independent pair of solutions. In contrast, $y_2 = \sin(t)$ and $y_3 = 17\sin(t)$ are called a *linearly dependent* pair of solutions.

Example. To get used to these ideas, answer the following questions which are all about the DE

$$(1) \qquad\qquad y'' + y = 0 .$$

1.　Show that $y_1 = \cos(t)$ and $y_2 = \sin(t)$ are solutions of the given DE by plugging them in.

2.　Show that $y_3 = 17\sin(t)$ is a solution of the same equation (1). Is there anything special about the constant 17 , or would any constant "work"?

Just how many solutions does a second order linear homogeneous equation have, and what exactly do they look like? To gain an insight into this question, do the following problems. Your CAS might be helpful here to do the algebra.

3.　Show that $y_4 = \cos(t) + \sin(t)$ and $y_5 = \cos(t) - \sin(t)$ are solutions of the DE (1). Are they a linearly independent or a linearly dependent pair?

4.　Repeat Question 3 for $y_6 = -\pi \cdot \cos(t) + i \cdot \sin(t)$ and $y_7 = -3 \cdot \cos(t) - 2.5 \cdot \sin(t)$.

5.　Find another solution (there's more than one), call it y_8 , such that y_8 and y_7 are a linearly *dependent* pair of solutions.

6. Show that, if $z_1(t)$ and $z_2(t)$ are <u>any</u> two solutions of $y'' + y = 0$, then
$z(t) = C_1 z_1(t) + C_2 z_2(t)$, where C_1 and C_2 are any constants, is also a solution. Do this in
general, without benefit of sines and cosines. This result says that any linear combination of
solutions is also a solution.

Question 6 illustrates the fact that the DE $y'' + y = 0$ has infinitely many solutions,
because choosing different values of the constants C_1 and C_2 will give different solutions. This
fact is true not only for this particular equation, but for any second order linear homogeneous
equation. The ***general second order linear homogeneous differential equation*** has the form:

(2)
$$a_0(t)y'' + a_1(t)y' + a_2(t)y = 0 ,$$

where the coefficients a_2, a_1, and a_0 are (usually continuous) functions of the independent
variable t. So far, we've been working with the case where these coefficients are <u>constant</u>
functions.

Is it possible to know what <u>all</u> the solutions must look like? The answer is yes! As it turns
out, if z_1 and z_2 are a <u>linearly independent pair</u> of solutions of the DE (2), then <u>every</u> other
solution can be written as a linear combination of z_1 and z_2. In symbols, this says that if $z(t)$ is
any old solution, then $z(t)$ must be identical to $C_1 z_1(t) + C_2 z_2(t)$ for some choice of the
constants C_1 and C_2. This is an extremely powerful result which you'll now be using.

- **Solving initial value problems, with and without your CAS**

When $y_1(t)$ and $y_2(t)$ are a linearly independent pair of solutions of

(1)
$$a_0(t)y'' + a_1(t)y' + a_2(t)y = 0 ,$$

then ***the general solution*** of equation (1) is

(2)
$$y(t) = C_1 y_1(t) + C_2 y_2(t) .$$

This means that <u>every</u> solution can be written as a linear combination of $y_1(t)$ and $y_2(t)$. There are no other kinds of solutions. You have the freedom to choose the constants C_1 and C_2.

Task 1. In previous lessons (see Lessons 23 and 25), you found two linearly independent solutions of each equation below. Now,
    a. find the general solution, by hand;
    b. then use your CAS to get a general solution. (See the Appendix for the syntax.)

    1. $\quad y'' - 6y' + 8y = 0$

    2. $\quad y'' - 4y' + 4y = 0$

    3. $\quad y'' - 6y' + 13y = 0$

You can get a <u>unique</u> solution by specifying the values of C_1 and C_2. One way to do this is to impose ***initial conditions*** on the solution, that is, specify the values that $y(t_0)$ and $y'(t_0)$ must take on, where t_0 is some point in the domain of $y(t)$. These values will then determine the constants C_1 and C_2. (You'll recall that in the spring equation you needed the initial values y_0 and v_0 in order to get a unique solution.) Step through the following example to see how this method works.

Example. Solve the *initial value problem* (IVP)

$$y'' - 6y' + 8y = 0 \, , \qquad y(0) = 1 \, , \quad y'(0) = 3 \, .$$

1. Write the characteristic equation and solve it.

2. Let $y_1(t)$ and $y_2(t)$ be the two linearly independent solutions you get from the eigenvalues. What are these solutions?

3. Write the general solution $y(t)$, and then find $y'(t)$.

4. Use the initial conditions and the general solution to form a system of two algebraic equations in the two unknowns C_1 and C_2. Here, $t_0 = 0$.

5. In Step 4 you should have found the system

$$y(0) = C_1 + C_2 = 1$$
$$y'(0) = 2C_1 + 4C_2 = 3$$

 Solve this system simultaneously for C_1 and C_2.

6. Now, given these values for the constants, write the resulting particular solution $y(t)$ of the IVP. You should end up with

$$y(t) = \frac{1}{2}e^{2t} + \frac{1}{2}e^{4t} \, .$$

 Check it: Is it a solution? Does it satisfy the initial conditions?

7. Use your CAS to solve the IVP.

Task 2. For each IVP below,
      a. solve it "by hand";
      b. then use your CAS to check your result.

$$1. \quad y'' + 4y = 0 \, , \quad y(\pi) = 1 \, , \quad y'(\pi) = 2$$
$$2. \quad y'' - y' = 0 \, , \quad y(0) = 0 \, , \quad y'(0) = 3$$

- **The mass-spring system revisited: Positive damping**

So far in our mass-spring computer experiment we have ignored the damping term by' in the differential equation. Now you'll explore the effect of a positive damping factor.

Task 1. Call up your 'spring' routine on your CAS (see Lessons 21 and 22.) Look again at the basic system spring(1,0,1,1,0) for $m = k = 1$, $y_0 = 1$, $v_0 = 0$, and damping factor $b = 0$.

Task 2. Before you see any more pictures, draw what you think the graph will look like when you increase b.

Task 3. As you did in Lesson 22, define a function of the single variable b, and get pictures of the position function, y(t), for the following values of b. Save your graphs for further analysis.

a) $b = 0.1$	d) $b = 1$	g) $b = 2$
b) $b = 0.5$	e) $b = 1.5$	h) $b = 2.2$
c) $b = 0.8$	f) $b = 1.9$	i) $b = 3$

Task 4. Identify any apparent equilibrium values in the above graphs. Which graph seems to achieve and remain at equilibrium value the most quickly?

Task 5. Speculate on what the graph will look like with a <u>negative</u> damping factor. Check your guess with the computer. Can you think of a reasonable interpretation for a negative damping factor in a "real" mass-spring system?

- ### The mass-spring system: Analysis of damping

Now you'll analyze the data you obtained in Lesson 28.

__Task 1.__ The DE you investigated in Lesson 28 is:

$$y'' + by' + y = 0.$$

Find the eigenvalues for this DE, with b as a parameter.

__Task 2.__ Find a value of b from your work in Lesson 28 where the picture seemed to change. Then use your solution of Task 1 above to explain why the change in behavior occurred at that particular value of b.

__Task 3.__ Continue solving the DE in Task 1, and write the general solution. You'll have to consider the size of b when writing the solution.

__Task 4.__ Use what you learned from Lesson 28 and the above analysis to make up reasonable definitions for the following terms:

a) an <u>underdamped</u> mass-spring system;

b) an <u>overdamped</u> mass-spring system;

c) a <u>critically damped</u> mass-spring system.

__Task 5.__ Use your definitions from Task 4 to classify the values of b from Lesson 28 as underdamped, overdamped, or critically damped.

- **Linear operators**

Look at the following DEs which you solved in Lesson 23:

$$y'' - 6y' + 8y = 0$$
$$y'' + 2y' - y = 0$$

These equations are classified as

second order, linear, __homogeneous__ .

Homogeneous means that $y(t) = 0$ is a solution. Another way of looking at it is that when you put the terms involving y and its derivatives on the left-hand side of the equation, the right-hand side is zero.

In contrast, look at this next pair of DEs:

$$y'' - 6y' + 8y = 10$$
$$y'' + 2y' - y = 3\sin 3t$$

These equations are classified as

second order, linear, __nonhomogeneous__ .

They're *nonhomogeneous* because $y(t) = 0$ is <u>not</u> a solution (or, the right-hand sides of the equations are not zero).

In order to solve nonhomogeneous equations, we must first investigate linearity, which is what we shall do now.

First, an *operator* is a function whose inputs are functions. For example, differentiation is an operator: if you put the funtion $f(t) = t^2$ into the derivative operator, the output is the function $g(t) = 2t$.

The left-hand side of each of the above DEs is an example of an operator. Moreoever, each is a <u>linear</u> operator. What that means will become clearer as you work along.

Task 1. Consider the following DE:

$$y'' + y' - 2y = 0 .$$

The <u>operator</u> is $L(y) = y'' + y' - 2y$. Solve this equation. Your solutions satisfy the **operator equation** $L(y) = 0$. Now use the solutions to fill in the first two entries in the table following Task 3 below.

Task 2. Filling in the rest of the table "by hand" can get tiresome, so we'll let your CAS do the work. See the Appendix on how to define and use an operator. Then use the computer to check your answers to Task 1 by defining the functions fa and fb as shown in the first two entries in the table and applying the operator L (defined in Task 1) to them, i.e., finding L(fa) and L(fb).

Task 3. Now use your CAS and the same operator L to fill in the rest of the table. The symbols C_1, C_2, a, b denote unknown constants.

<u>y</u>	<u>L(y)</u>
$fa(t) = e^t$	
$fb(t) = e^{-2t}$	
$fc(t) = C_1 \cdot e^t + C_2 \cdot e^{-2t}$	
$fd(t) = e^{3t}$	
$fe(t) = a \cdot e^{3t}$	
$ff(t) = e^{4t}$	
$fg(t) = a \cdot e^{3t} + b \cdot e^{4t}$	
$fh(t) = \sin(t)$	
$fj(t) = a \cdot \sin(t)$	
$fk(t) = \cos(t)$	
$fl(t) = a \cdot \sin(t) + b \cdot \cos(t)$	
$fm(t) = \cos(2t)$	
$fn(t) = a \cdot \cos(2t) + b \cdot \sin(2t)$	

Task 4. Are the following true or false?

________ $L(a{\cdot}y) = a{\cdot}L(y)$

________ $L(y_1 + y_2) = L(y_1) + L(y_2)$

Any operator that satisfies these two properties is called a ***linear operator***.

Task 5. Now do four problems "backwards". Fill in the table:

y	$\underline{L(y)}$
fo(t) =	$3e^{-t}$
fp(t) =	$e^{-t} + 2e^{5t}$
fq(t) =	$\sin(t)$
fr(t) =	$2\cos(3t) - \sin(3t)$

- **Solving a nonhomogeneous equation: Undetermined coefficients**

The idea behind solving a nonhomogeneous equation is to look at the right-hand side of the DE and to guess what kind of function went into the linear operator to cause the right-hand side to come out. That is, the procedure is similar to Task 5 of Lesson 30. Let's do some examples.

Example 1. Find a solution of the nonhomogeneous equation $y'' + 3y' + 2y = e^t$.

Here we have the linear operator $L(y) = y'' + 3y' + 2y$ and we want a function $y(t)$ so that $L(y(t)) = e^t$ on some t-interval.

Since the right-hand side function e^t is its own first and second derivative, we might try $y(t) = Ae^t$ as the solution we're seeking, where A is some constant to be found. We plug this in to get

$$\begin{aligned} L(Ae^t) &= (Ae^t)'' + 3(Ae^t)' + 2(Ae^t) \\ &= Ae^t + 3Ae^t + 2Ae^t \\ &= (A + 3A + 2A)e^t \\ &= 6Ae^t. \end{aligned}$$

Because we want this last function to be identical to e^t for all t on some interval, we must have

$$6Ae^t = e^t \qquad \text{or} \qquad 6A = 1.$$

This gives $A = 1/6$ so the particular solution we want is $y(t) = (1/6)e^t$. Plug it in to make sure.

This solution technique is called the ***method of undetermined coefficients,*** and it boils down to solving for constants like A in the above example.

WARNING: The method of undetermined coefficients only works when the DE is linear with constant coefficients and the right-hand side of the DE consists of functions whose successive derivatives "repeat themselves", like exponential, sine, and cosine functions. It will also work if the right-hand side is a polynomial, since the derivatives of polynomials are other polynomials. The method even works if the right-hand side of the DE is a sum of products of such functions. It won't work when the right-hand side is something like $\ln(t)$ or $\csc(t)$.

Let's do another example.

Example 2. Find a solution of $y'' + 3y' + 2y = t^2$.

We have the same linear operator L as in Example 1, but now we want a function $y(t)$ so that $L(y(t)) = t^2$. From looking at the right-hand side of the DE, we might try a solution $y(t) = At^2$. We have

$$
\begin{aligned}
L(y(t)) &= L(At^2) \\
&= (At^2)'' + 3(At^2)' + 2(At^2) \\
&= 2A + 3(2At) + 2(At^2) \\
&= 2A + 6At + 2At^2 ,
\end{aligned}
$$

and we want this last expression to be identical to t^2 . Two polynomials are identical if and only if they have the same coefficients, and this means that A would have to equal both 0 and $\tfrac{1}{2}$, which is impossible. So our trial solution failed.

The problem is that $L(y)$ involves y' and y'' too, so we need to include lower degree t-terms in our trial solution. Let's try

$$
y(t) = At^2 + Bt + C .
$$

Plug this in to get (do the details):

$$
(*) \qquad L(y(t)) = 2At^2 + (6A + 2B)t + (2A + 3B + 2C) .
$$

Now, the right side of the DE is t^2 , or, more fully, $1 \cdot t^2 + 0 \cdot t + 0$, and this will equal the right side of (*) if and only if the coefficients are identical. So $A, B,$ and C must satisfy the equations

$$
\begin{aligned}
2A &= 1 , \\
6A + 2B &= 0 , \\
2A + 3B + 2C &= 0 .
\end{aligned}
$$

The solution of this sytem is $A = 1/2$, $B = -3/2$, $C = 7/4$ (check this), so the solution of the DE we want is

$$
y(t) = (1/2)t^2 - (3/2)t + (7/4) .
$$

Check that this is a solution by plugging it into the DE.

Task 1. Find a solution of $y'' + 3y' + 2y = t^3 + 7$, and check your answer.

Task 2. Find a solution of the nonhomogeneous DE $y'' + 3y' + 2y = \sin(t)$.
Should you try just $y(t) = A\sin(t)$ or should you include more in your trial solution?

Task 3. Apply the "educated guessing" method of undetermined coefficients to find at least one solution for each of the following nonhomogeneous DEs: (This is the same problem as Task 5 of Lesson 30, just stated differently.) See the Appendix for some helpful computer techniques.

1. $y'' + y' - 2y = 3e^{-t}$

2. $y'' + y' - 2y = e^{-t} + 2e^{5t}$

3. $y'' + y' - 2y = \sin(t)$

4. $y'' + y' - 2y = 2\cos(3t) - \sin(3t)$

Task 4. What happens when you use this method to try to find a particular solution of

$$L(y) = y'' + y' - 2y = e^{t} ?$$

What is your next best guess for a trial solution? (Hint: Recall what the solutions are when you have a repeated real eigenvalue in the homogeneous case.)

Task 5. Obviously, the method needs adjusting when the right-hand function in the DE is itself a solution of the associated homogeneous equation. Can you generalize your result in Task 4? That is, complete the following sentence:

If you want a particular solution of the nonhomogeneous DE $L(y) = f(t)$, and if $f(t)$ is already a solution of the associated homogeneous DE $L(y) = 0$, then your trial solution should be

_______________________.

Task 6. Find a particular solution of $y'' + y = \sin(t)$. (Hint: First solve the associated homogeneous DE $y'' + y = 0$ so you'll know how to construct a trial solution; see Task 5.)

- **The general solution of the linear nonhomogeneous equation**

Think of a typical second order linear <u>homogeneous</u> DE, say the mass-spring equation

(*spring*) $$my'' + by' + ky = 0 \ .$$

You know that the general solution is of the form $C_1y_1(t) + C_2y_2(t)$. To get a specific solution, you are free to choose the values of C_1 and C_2 arbitrarily; we sometimes express this by saying that there are <u>two degrees of freedom</u>.

So far, the solutions of our <u>nonhomogeneous</u> equations in Lesson 31 have had no arbitrary constants at all. We have not found their general solutions, only particular solutions. Let's try to figure out what the general solution might look like. Look at the familiar equation from Lesson 31, Task 3.1:

$$L(y) = y'' + y' - 2y = 3e^{-t} \ .$$

1. Retrieve your particular solution to this nonhomogeneous equation. Call this solution y_p. What is $L(y_p)$?

2. Try multiplying this solution by a constant k. Test the new function to see whether it, too, is a solution. (Oops!)

3. Now find the general solution to the <u>associated homogeneous</u> equation

$$L(y) = y'' + y' - 2y = 0 \ .$$

Call this solution y_h. What is $L(y_h)$?

4. Use the above answers and the linearity principle from Lesson 30, Task 4 to predict the result of $L(y_p + y_h)$. Check your prediction.

5. The <u>general solution</u> of the nonhomogeneous equation $L(y) = 3e^{-t}$ must have two properties: it must be a function which satisfies the DE and it must have two degrees of freedom. You are now in possession of this general solution. What is it?

You have just discovered another major result, which can be stated as follows:

If $y_h(t) = C_1 y_1(t) + C_2 y_2(t)$ is the general solution of the <u>homogeneous</u> DE

(1) $$a_0(t)y'' + a_1(t)y' + a_2(t)y = 0 ,$$

and if $y_p(t)$ is a <u>particular solution</u> of the <u>nonhomogeneous</u> DE

(2) $$a_0(t)y'' + a_1(t)y' + a_2(t)y = f(t) ,$$

then <u>every</u> solution of (2) is of the form

$$y_p(t) + y_h(t) .$$

The function $y_p(t) + y_h(t)$ is called the **general solution** of the nonhomogenerous DE (2).

Task 1. Retrieve your particular solutions to Task 3 of Lesson 31 and write the general solutions of these nonhomogeneous DEs.

Task 2. Write the general solutions for the equations in Tasks 4 and 6 of Lesson 31.

Task 3. Use your CAS to find general solutions of the nonhomogeneous DEs in the above Tasks 1 and 2. (Just use the 'dsolve' command in Maple, or the 'DSolve' command in Mathematica.) Do your answers agree with the computer's? You might have to do some algebraic juggling to make them agree.

Task 4. Now practice solving a few initial value problems. You learned how to do these for homogeneous equations in Lesson 27; the technique is the same here. Do these by hand first, then use your CAS to solve them.

1. $y'' + y' - 2y = 3e^{-t} , \qquad y(0) = 1 , \quad y'(0) = 2$
2. $y'' + y' - 2y = \sin(t) , \qquad y(\pi/2) = 0 , \quad y'(\pi/2) = 1$
3. $y'' + 4y = t^2 , \quad y(0) = -1 , \quad y'(0) = -2$

- **Forced mass-spring systems**

Recall the mass-spring system DE,

$$(1) \qquad my'' + by' + ky = 0 \, ,$$

where m, b, and k are positive constants.

When $b = 0$ there is no friction. The resulting equation is often written as $y'' + \omega^2 y = 0$ and is called a ***harmonic oscillator equation,*** a reference to its application in mechanics and electrical circuits. The motion of its sine and cosine solutions is called ***simple harmonic motion.***

If you apply an external ***forcing function, f(t)***, to the mechanical or electrical system, you get the following nonhomogeneous equation,

$$(2) \qquad my'' + by' + ky = f(t) \, ,$$

called a ***forced harmonic oscillator equation.*** Some of the forcing functions of interest are various sine and cosine functions. You'll investigate some of these equations now.

Task 1. In this task, think of $f(t)$ as a periodic force applied to a mass-spring system. Find the particular solutions for each of the following IVPs. Use the same initial values for each equation, namely, $y(0) = 1$ and $y'(0) = 0$, so that you can compare results.

1. $y'' + y = 0 \qquad$ (*unforced*)

2. $y'' + y = \sin(2t)$

3. $y'' + y = 3\sin\!\left(\dfrac{t}{3}\right)$

4. $y'' + y = 2\sin\!\left(\dfrac{3t}{5} + \dfrac{\pi}{4}\right)$

Task 2. For each of the solutions in Task 1,
1. Give the period and the maximum displacement from equilibrium.
2. Compare and contrast the graphs of these four solutions. For example, which solutions are bounded? How does the coefficient of t in the forcing function affect the solution?

Task 3. In Lesson 31, Task 6, you were asked to find a particular solution of the DE

$$y'' + y = \sin(t) .$$

The method of undetermined coefficients gives the solution

$$y_p(t) = -\tfrac{1}{2}\, t \cos(t) ,$$

which is obtained from the trial solution $y_p = A\, t \sin(t) + B\, t \cos(t)$. (The extra t factor was added because the forcing function, $f(t) = \sin(t)$, is itself a solution of the associated homogeneous DE.)

1. Use your CAS to solve the IVP $y'' + y = \sin(t)$, $y(0) = 1$, $y'(0) = 0$.

2. Look at the graph of the solution and describe its behavior. This behavior is called *resonance*.

Task 4. Suppose you have an undamped spring of spring constant $k = 1$ and you apply an outside forcing function $f(t) = \sin(2t)$. Find a mass m that will make the system resonate. Beginning at equilibrium with velocity 0 , how soon will the amplitude of the system exceed 10 ?

• **More about linear operators**

Recall that an operator is a <u>function</u> whose inputs and outputs are <u>functions</u>, rather than numbers. In Lesson 30 we defined the differential operator

$$L(y) = y'' + y' - 2y$$

and you used your CAS in Task 3 to deduce that the operator L satisfies the following two properties:

(1) $$L(a \cdot y) = a \cdot L(y) ,$$

(2) $$L(y_1 + y_2) = L(y_1) + L(y_2) ,$$

where a is any constant and y, y_1, y_2 are any functions of t. Any operator that satisfies properties (1) and (2) is called a ***linear operator.***

Task 1. To illustrate properties (1) and (2) above, let $L(y) = y'' + y' - 2y$ and take $y(t) = e^{3t}$.

1. Calculate (by hand) the left and right sides of property (1) for this L. (You should get $4a \cdot e^{3t}$ both times.)

2. Now take $y_1 = \sin(t)$ and $y_2 = e^{3t}$ and illustrate property (2) by hand.

Now you can practice identifying linear operators.

Task 2. Determine which of the following differential operators is linear, and which is not (these are called ***nonlinear***). For the nonlinear ones, tell which of the two linearity properties fails to hold.

1. $R(y) = y'' - y' + y^2$
2. $S(y) = y'y'' - 2y$
3. $T(y) = y' - 2ty$
4. $V(y) = t\, y'' - 2t^{\frac{1}{2}}\, y' + y \sin(t)$

In Task 2, R and S are nonlinear, while T and V are both linear. Are you surprised? Remember that these operators are functions of y, not t.

Task 3. Write two more examples each of linear differential operators and of nonlinear ones.

In the study of differential equations, it is very important that you be able to distinguish between linear and nonlinear equations. A *linear equation* is a DE of the form $L(y) = f(t)$, where L is a linear differential operator. A *nonlinear equation* is any other kind of DE. The theory of linear DEs has been well-developed over the past 150 years or so; in contrast, the theory of nonlinear equations is not an easy nut to crack. Some of them lead to chaotic behavior, and we are only beginning to understand these with the help of today's powerful computers.

The ***general second order linear differential operator*** is

$$G(y) = a_0(t)y'' + a_1(t)y' + a_2(t)y \, ,$$

and so the ***general second order linear differential equation*** is of the form $G(y) = f(t)$, or

$$(3) \qquad\qquad a_0(t)y'' + a_1(t)y' + a_2(t)y = f(t) \, .$$

If equation (3) is <u>homogeneous</u>, that is, $f(t) \equiv 0$ for all t, then you saw in Question 6 of Lesson 26 that if $z_1(t)$ and $z_2(t)$ are any two solutions, then the linear combination $C_1 z_1(t) + C_2 z_2(t)$ is also a solution. You are now able to prove this result, using the properties of linear operators.

<u>**Task 5.**</u> Here's the problem statement: Given that $z_1(t)$ and $z_2(t)$ are solutions of the linear homogeneous DE

$$(4) \qquad\qquad G(y) = a_0(t)y'' + a_1(t)y' + a_2(t)y = 0 \, ,$$

show that $C_1 z_1(t) + C_2 z_2(t)$ is also a solution. To prove this, first notice that $G(z_1(t)) = 0$ and $G(z_2(t)) = 0$ (why?). Now give reasons for each of the following steps in the rest of the proof:

$$
\begin{aligned}
G(C_1 {\cdot} z_1(t) + C_2 {\cdot} z_2(t)) &= G(C_1 {\cdot} z_1(t)) + G(C_2 {\cdot} z_2(t)) \\
&= C_1 {\cdot} G(z_1(t)) + C_2 {\cdot} G(z_2(t)) \\
&= C_1 {\cdot} 0 + C_2 {\cdot} 0 \\
&= 0 \, ,
\end{aligned}
$$

thus showing that a linear combination of solutions of the homogeneous DE is also a solution.

<u>**Task 6.**</u> Show that the result of Task 5 is <u>not</u> true for <u>nonhomogeneous</u> linear DEs by finding what's called a ***counterexample,*** that is, find a <u>specific</u> nonhomogeneous DE: $G(y) = f(t)$ and any two of its solutions, say $u_1(t)$ and $u_2(t)$, and show that $G(C_1 u_1(t) + C_2 u_2(t)) \neq f(t)$, that is, a linear combination of solutions is <u>not</u> a solution.

● The wronskian

Suppose you were going to solve the IVP

$$y'' + y' - 2y = 0 , \quad y(0) = 1 , \quad y'(0) = 2 .$$

You would find the general solution

$$y(t) = C_1 e^t + C_2 e^{-2t}$$

and then use the initial conditions to set up the following system of algebraic equations

$$\text{(1)} \qquad \begin{aligned} y(0) &= C_1 + C_2 = 1 \\ y'(0) &= C_1 - 2C_2 = 2 \end{aligned}$$

and proceed to solve for the constants C_1 and C_2. Do so.

How did you know you'd get a solution? Think of the geometry of the situation: in a plane with axes labelled C_1 and C_2, the equations of system (1) represent two lines which intersect at the solution point $(4/3 , -1/3)$. However, if the lines happened to be parallel, there would be no solution, which would mean the original IVP would not have a solution. And if the lines were coincident, there would be infinitely many solutions, which would imply infinitely many solutions of the IVP. Actually, the Existence and Uniqueness Theorem (see your text) tells us that this initial value problem does have a solution, and, furthermore, there is only one.

To gain an insight into this problem, consider the general case. If you can find two linearly independent solutions $y_1(t)$ and $y_2(t)$, then you can solve the second order linear homogeneous IVP

$$\text{(2)} \qquad a_0(t)y'' + a_1(t)y' + a_2(t)y = 0 , \quad y(t_0) = y_0 , \quad y'(t_0) = v_0$$

by doing the following steps:

1. Write the general solution of the DE in (2); call this solution $y(t)$.

2. You then want to find C_1 and C_2 by solving the following algebraic system:

$$\text{(3)} \qquad \begin{aligned} y(t_0) &= C_1 y_1(t_0) + C_2 y_2(t_0) = y_0 \\ y'(t_0) &= C_1 y_1'(t_0) + C_2 y_2'(t_0) = v_0 . \end{aligned}$$

(Don't be put off by the notation: remember that $y_1(t_0)$, $y_2(t_0)$, etc., are just constants with fancy names; the variables here are C_1 and C_2.) To solve the system (3), you can use the method of elimination: multiply the first equation by $y_1{}'(t_0)$ and the second equation by $y_1(t_0)$ and subtract the equations. You should get a single equation in the unknown C_2. Solve it for C_2 and write your answer as a single fraction over a common denominator.

3. Now solve equations (3) for C_1 in a similar fashion.

4. Both denominators in your answers for C_1 and C_2 should be the same, and should be equal to:

$$W(t_0) = y_1(t_0)\, y_2'(t_0) - y_1'(t_0)\, y_2(t_0) .$$

$W(t)$ is called the **wronskian** of the solutions y_1 and y_2. What becomes of C_1 and C_2 if $W(t_0) = 0$?

Now you'll investigate the wronskian.

Task 1. Below are some homogeneous DEs you solved earlier, along with two solutions for each one. Compute the wronskian, $W(t)$, for each pair of solutions. Your CAS will be handy here.

1. $y'' + y = 0$, $\quad y_1 = \sin(t)$, $\quad y_2 = \cos(t)$

2. $y'' + y' - 2y = 0$, $\quad y_1 = e^t$, $\quad y_2 = e^{-2t}$

3. $y'' + 10y' + 25y = 0$, $\quad y_1 = e^{-5t}$, $\quad y_2 = te^{-5t}$

Task 2. What are the zeros of $W(t)$ for each of the wronskians in Task 1? That is, for what t-values does the wronskian equal zero?

Task 3. Here are the same equations as in Task 1, but this time with a different pair of solutions. Find the wronskian for each pair of solutions.

1. $y'' + y = 0$, $\quad y_1 = \sin(t)$, $\quad y_2 = 3\sin(t)$

2. $y'' + y' - 2y = 0$, $\quad y_1 = e^t$, $\quad y_2 = -e^t$

3. $y'' + 10y' + 25y = 0$, $\quad y_1 = te^{-5t}$, $\quad y_2 = 2te^{-5t}$

Task 4. What are the zeros of the wronskians of Task 3?

Task 5. Compare your answers to Tasks 2 and 4. What do you observe? Can you think of a reason that will explain the difference in behavior?

You've just illustrated an important fact about wronskians, namely:

Two solutions $y_1(t)$ and $y_2(t)$ of the DE $a_0(t)y'' + a_1(t)y' + a_2(t)y = 0$ are linearly dependent if and only if their wronskian is identically zero.

It can also be shown that if the leading coefficient, $a_0(t)$, is never equal to zero on the interval you're working on, then the wronskian is either never zero or is identically zero there.

So the wronskian is a convenient testing device for linear dependence or independence of solutions of a linear homogeneous DE. This might not seem like a big deal for second order equations, because you can usually tell on sight if one solution is a constant multiple of another, but it becomes useful for higher order equations, when you have three, four, or more solutions to test for linear independence.

- **Solving a nonhomogeneous equation: Variation of parameters**

Now you'll learn a more general method than that of undetermined coefficients for finding a particular solution of a linear nonhomogeneous DE.

Example. Suppose that you wish to find a particular solution of the linear DE

(1) $$L(y) = y'' + y = csc(t) \,.$$

Using the method of undetermined coefficients that you learned in Lesson 31 would have you use $y_p(t) = A{\cdot}csc(t)$ as a trial solution and would result in:

$$L(A{\cdot}csc(t)) = A{\cdot}csc^3(t) + A{\cdot}csc(t)\cot^2(t) + A{\cdot}csc(t) \,,$$

which cannot be matched up to equal $csc(t)$ no matter what you choose for A. The problem is that consecutive derivatives of $csc(t)$ don't "repeat" themselves in any pattern, as do those of $exp(t)$, $sin(t)$, $cos(t)$, and polynomials.

The following method, called *variation of parameters,* will <u>always</u> work, even for the variable coefficient case, provided that you know the general solution of the associated homogeneous equation, $L(y) = 0$.

To learn the method, do the following steps.

1. Write the associated homogeneous equation for equation (1) above and solve it.

2. Now change the constants C_1 and C_2 of your answer in Step 1 into unknown <u>functions</u> $v_1(t)$ and $v_2(t)$. At this point you should have

(2) $$y_p(t) = v_1(t)\cos(t) + v_2(t)\sin(t) \,.$$

This function $y_p(t)$ will be your trial solution for equation (1). The idea is to determine the <u>functions</u> $v_1(t)$ and $v_2(t)$ so that $y_p(t)$ will satisfy equation (1). In the method of undetermined coefficients, you had to find constants; now you must find functions.

3. Let's start to plug $y_p(t)$ into the DE; so find $y_p{}'(t)$.

Your answer to Step 3 should be

$$(3) \qquad y_p'(t) = [-v_1(t)\sin(t) + v_2(t)\cos(t)] + [v_1'(t)\cos(t) + v_2'(t)\sin(t)] .$$

Study equation (3) and realize that the next derivative, $y_p''(t)$, will contain the factors $v_1''(t)$ and $v_2''(t)$, so that, when we plug our trial solution into the DE, we'll have the first and second derivatives of the <u>two</u> unknown functions in it. This does not look promising! But all is not lost.

Because we're looking for <u>two</u> unknown functions, we can impose <u>two</u> conditions (i.e., equations) on them. One of these conditions is, of course, the DE itself, namely, $L(y_p) = \csc(t)$. The other condition, we shall decide, is to require that the second term in brackets in equation (3) be identically zero, thus eliminating the troublesome $v_1'(t)$ and $v_2'(t)$ terms . So the second condition is

$$(4) \qquad v_1'(t)\cos(t) + v_2'(t)\sin(t) = 0 .$$

4. After imposing condition (4) on equation (3), find $y_p''(t)$.

5. Now plug $y_p(t)$ and its derivatives into the DE $L(y_p) = \csc(t)$, using your results from Steps 3 and 4.

The terms containing $v_1(t)$ and $v_2(t)$ will vanish from the calculations in Step 5 because $L(\cos(t)) = 0$ and $L(\sin(t)) = 0$ (why?). So you should end up with the simple equation

$$(5) \qquad L(y_p) = -v_1'(t)\sin(t) + v_2'(t)\cos(t) = \csc(t) .$$

6. The two equations, (4) and (5), are really just two <u>algebraic</u> equations in the two unknowns $v_1'(t)$ and $v_2'(t)$. Solve them simultaneously.

7. Did you get $v_1'(t) = -1$ and $v_2'(t) = \cot(t)$ in Step 6? Plug these answers back into equations (4) and (5) to make sure they work.

8. Integrate the functions in Step 7 to finally get $v_1(t)$ and $v_2(t)$. You may take the integration constants to be zero since you're looking for a single pair of functions.

9. Finally, you can now write the particular solution $y_p(t)$. Do it. Then plug it into the DE to make sure it works.

Your final answer should be

$$y_p(t) = -t\cos(t) + \sin(t) \cdot \ln(\sin(t)) \ .$$

Fortunately, you don't have to do all these calculations for every problem. All you need are equations (4) and (5).

> <u>Summary of the method:</u> To find a particular solution, $y_p(t)$, of the second order linear nonhomogeneous DE
>
> (1) $\quad L(y) = a_0(t)y'' + a_1(t)y' + a_2(t)y = f(t)$,
>
> do the following:
>
> a) Find two linearly independent solutions, $y_1(t)$ and $y_2(t)$ of the associated homogeneous DE $L(y) = 0$;
>
> b) Set up the following system of algebraic equations and solve for the unknowns $v_1{}'(t)$ and $v_2{}'(t)$:
>
> $$v_1'(t)y_1(t) + v_2'(t)y_2(t) = 0$$
>
> $$v_1'(t)y_1'(t) + v_2'(t)y_2'(t) = \frac{f(t)}{a_0(t)}$$
>
> c) Integrate $v_1{}'(t)$ and $v_2{}'(t)$ (if possible) to get $v_1(t)$ and $v_2(t)$.
>
> d) The particular solution is then
>
> $$y_p(t) = v_1(t)y_1(t) + v_2(t)y_2(t) \ .$$

<u>Task 1.</u> Use the method of variation of parameters outlined above to find a particular solution for each of the following DEs. Write the general solution, too. (Use your CAS to do some of the algebra and integration.)

1. $y'' - y' - 6y = e^{\,t}$
2. $y'' + 4y = \tan(2t)$
3. $y'' + 2y' + y = e^{-t}\ln(t)$

- **Euler equations: Distinct roots**

Here's a type of equation you haven't seen yet. It's linear but it has <u>variable</u> coefficients, and we can find closed form solutions for it. It's called an ***Euler equation***.

(1) $$t^2 y'' - 2ty' + 2y = 0 .$$

Look closely at its <u>form</u>. It's almost a linear combination of y, y', and y'' set equal to zero, but the coefficients are polynomial functions of t rather than constants. Can you guess what a solution might look like? (Hint: Look at the <u>powers of t</u> in the coefficients.)

What if we tried a solution of the form $y = t^r$, where r is a constant to be found. Let's do it.

1. Plug the trial solution $y = t^r$ into equation (1).

Your final equation in Step 1 should have a factor of t^r which you can divide out (assume $t \neq 0$) to get

(2) $$r(r - 1) - 2r + 2 = 0 ,$$

a simple quadratic equation in the variable r. This equation is called ***the indicial equation*** for the Euler DE (1) and is very much like the characteristic equation you saw in the constant coefficient case.

2. Solve the indicial equation (2) for the roots r_1 and r_2. You should get $r_1 = 1$ and $r_2 = 2$. Thus, two solutions of the DE (1) will be

$$y_1(t) = t \quad \text{and} \quad y_2(t) = t^2 .$$

Check that they are solutions by plugging them into the DE.

3. Are y_1 and y_2 linearly independent?

4. Write the general solution of equation (1). (Remember that equation (1) is a <u>linear</u> equation.)

Task 1. The general solution of equation (1) is $C_1 t + C_2 t^2$. Find the solution which satisfies the initial conditions $y(1) = 1$ and $y'(1) = -2$.

Task 2. Show that $L(y) = t^2 y'' - 2ty' + 2y$ is a linear operator. (See Lesson 34 for the definition of a linear operator.)

Task 3. Solve the following Euler equations by hand using the trial solution $y = t^r$. Check your answers by plugging the solutions back into the DEs.

1. $t^2 y'' - 3ty' + 3y = 0$
2. $t^2 y'' - 2y = 0$
3. $4t^2 y'' - 4ty' + 3y = 0$

Task 4. Now use your CAS to solve the equations of Task 3 directly.

Task 5. Study the <u>forms</u> of the equations in Task 3, and then write the <u>general form</u> of a second order Euler equation. Can you write the general form of a third order Euler equation?

Task 6. Are the following equations Euler equations? If not, why not?

1. $t^2 y'' + 2t^2 y' + 3y = 0$
2. $t^2 y'' + (t + 1)y' - 2y = 0$
3. $5t^2 y'' + 7ty' - 3ty = 0$

Task 7. Here's the ***general second order Euler equation***. Find its indicial equation and the roots in terms of the coefficients a, b, c.

$$at^2 y'' + bty' + cy = 0, \quad \text{where } a, \ b, \ c \text{ are constants.}$$

● **Euler equations: Repeated roots and reduction of order**

When the indicial polynomial of an Euler equation has a repeated root, we'll only get one solution from it, and multiplying that solution by t doesn't give us another solution, as it did in the constant coefficient case. Verify these comments by doing the following two steps.

1. Show that the Euler equation

(1) $$t^2 y'' - 3ty' + 4y = 0$$

has the repeated root $r = 2$, thus giving us only one solution $y_1(t) = t^2$.

2. Show that multiplying the solution $y_1(t)$ by t does <u>not</u> give another solution of the DE (1).

Now what shall we do to get a second, linearly independent solution? There is a standard way of finding another linearly independent solution if one solution is already known. It's called the *reduction of order method.* Continue with this example to discover how the method works.

3 Since multiplying $y_1(t) = t^2$ by t didn't give another solution, we'll try multiplying it by a <u>function</u> v(t) and see if we can discover a v(t) that will make it work. So let

$$y_{trial}(t) = t^2 \cdot v(t) .$$

Plug $y_{trial}(t)$ into the DE (1).

You should now have a new DE in the variable v , namely, the DE

(2) $$t^4 v'' + t^3 v' = 0 ,$$

or

(3) $$tv'' + v' = 0 , \qquad t \neq 0 .$$

Look closely at equation (3): there is no v-term, so we can think of (3) as a <u>first</u> order DE in the new variable $w = v'$. (This is why the method is called "reduction of order".) Let's make the substitution $w = v'$ with $w' = v''$ in (3) and rearrange it slightly to get (do the steps):

(4) $$w' + \frac{1}{t}w = 0 , \quad t \neq 0 .$$

4. Equation (4) is both linear and separable, so you know how to solve it. Do so, assuming
   that $t > 0$.

   The solution of equation (4) is $w = C/t$ for $t > 0$, and we can now work backwards to
get v. Since we let $v' = w$, we can integrate this to get

$$v = \int \frac{C}{t}\, dt = C\ln(t) + C_1, \quad \textit{for} \quad t > 0 .$$

Any one of these functions will do as the function v we're after (except $v = 0$, of course), so
choose $C = 1$ and $C_1 = 0$ and use $v = \ln(t)$. Our desired second solution of the original Euler
equation (1) is then

$$y_2(t) = t^2 \cdot v(t) = t^2 \ln(t) , \quad \textit{for} \quad t > 0 .$$

5. Plug $y_2(t)$ into the DE (1) to convince yourself that it is a solution.

6. Are the two solutions y_1 and y_2 a linearly independent pair? What is the general solution?

Task 1. Find a second solution for equation (1), this time assuming that $t < 0$. Check your
answer.

Task 2. Use the above reduction of order method to find the general solution of the Euler DE

$$t^2 y'' - ty' + y = 0 .$$

Check your answer by plugging it into the DE, and also solve the DE directly with your CAS.

   The beauty of the reduction of order method is that it works for <u>any</u> linear DE of the form

$$y'' + p(t)y' + q(t)y = 0 ,$$

not just for Euler equations.

Task 3. Use the reduction of order method to find a second linearly independent solution for
each equation below, then see if your CAS can solve them. A first solution, $y_1(t)$, is given for
each. Notice that these are not Euler equations, but they are linear.

1. $ty'' - (t+1)y' + y = 0$, for $t > 0$; $y_1(t) = e^t$

2. $ty'' + (1 - 2t)y' + (t-1)y = 0$, for $t > 0$; $y_1(t) = e^t$

●   **Euler equations: Complex roots and changing the independent variable**

Let's look at the relationship between an Euler DE and a constant coefficient equation: both have a companion polynomial whose roots determine the solutions of the DE. There's too much similarity here for us to ignore.

Consider the Euler equation

(E1) $$t^2 \frac{d^2y}{dt^2} - 2t\frac{dy}{dt} + 2y = 0$$

from Lesson 37. It has the indicial equation (verify this)

(P1) $$r^2 - 3r + 2 = 0 \, ,$$

which has roots $r_1 = 1$ and $r_2 = 2$, so that equation (E1) has the pair of linearly independent solutions

$$y_1(t) = t \qquad and \qquad y_2(t) = t^2 \, .$$

But we can also view the indicial equation (P1) as the <u>characteristic equation</u> of the constant coefficient DE

(CC1) $$\frac{d^2z}{dx^2} - 3\frac{dz}{dx} + 2z = 0 \, ,$$

which has solutions

$$z_1(x) = e^x \qquad and \qquad z_2(x) = e^{2x} \, .$$

We've written the DEs (E1) and (CC1) using different variable names so that you can better see what's going on in what follows.

The two DEs, equations (E1) and (CC1), have the same polynomial equation (P1). How are their solutions $y_1(t) = t$ and $z_1(x) = e^x$ related? The first one is a power function and the second one is an exponential function, but if we substitute $t = e^x$ in $y_1(t)$, we'll get $z_1(x)$. Likewise, the solution $y_2(t) = t^2$ will become $z_2(x) = e^{2x}$, since $t^2 = (e^x)^2 = e^{2x}$.

The same thing happens in a repeated root case. This evidence leads us to suspect that letting $t = e^x$ in any Euler DE will reduce it to a constant coefficient equation.

To see how this works, let's use equation (E1). To carry out the substitution $t = e^x$, we need to express dy/dt in terms of dz/dx and d^2y/dt^2 in terms of d^2z/dx^2, and we can do this with the chain rule.

The chain rule gives

(1)
$$\frac{dy}{dt} = \frac{dz}{dx} \cdot \frac{dx}{dt} \ .$$

Let's verify (1) for the solutions $y_2(t) = t^2$ and $z_2(x) = e^{2x}$. If $y_2(t) = t^2$ and we let $t = e^x$, then the left-hand side of (1) is

$$\frac{dy_2}{dt} = 2t \ .$$

For the right-hand side, we'll need dx/dt. Since $t = e^x$, then $x = \ln(t)$, and so $dx/dt = 1/t$. The right side of (1) now becomes

$$\frac{dz_2}{dx} \cdot \frac{dx}{dt} = 2e^{2x} \cdot \frac{1}{t} \ , \qquad \text{since} \quad \frac{dx}{dt} = \frac{1}{t} \ ,$$

$$= 2e^{2x} \cdot \frac{1}{e^x} \ , \qquad \text{since} \quad t = e^x \ ,$$

$$= 2e^x$$

$$= 2t \ .$$

Thus, both sides of (1) reduce to the same function of t.

To get the relationship between the second derivatives, rewrite (1), using the fact that $dx/dt = 1/t$:

(2)
$$\frac{dy}{dt} = \frac{dz}{dx} \cdot \frac{1}{t} \ .$$

Keep in mind that dz/dx is a <u>function of</u> t because of the relationship $x = \ln t$. Now differentiate both sides of (2) <u>with respect to</u> t and use the product rule to get

$$\frac{d^2y}{dt^2} = \frac{d}{dt}\left(\frac{dz}{dx}\cdot\frac{1}{t}\right)$$

$$(3) \qquad = \frac{d}{dt}\left(\frac{dz}{dx}\right)\cdot\left(\frac{1}{t}\right) + \frac{dz}{dx}\cdot\left(-\frac{1}{t^2}\right) .$$

The factor $\dfrac{d}{dt}\left(\dfrac{dz}{dx}\right)$ in the first term on the right of (3) can be rewritten using the chain rule:

$$\frac{d}{dt}\left(\frac{dz}{dx}\right) = \frac{d}{dx}\left(\frac{dz}{dx}\right)\cdot\frac{dx}{dt}$$

$$(4) \qquad = \frac{d^2z}{dx^2}\cdot\frac{1}{t} , \quad \text{since } \frac{dx}{dt} = \frac{1}{t} ,$$

and substituting (4) into the first term on the right side of (3) finally gives

$$(5) \qquad \frac{d^2y}{dt^2} = \frac{d^2z}{dx^2}\cdot\frac{1}{t^2} - \frac{dz}{dx}\cdot\frac{1}{t^2} .$$

(That was a little tricky. Go back over it a few times to convince yourself of its validity.)

Finally, we can substitute equations (5) and (2) into the Euler DE (E1) and watch it turn into the constant coefficient DE (CC1):

$$(6) \qquad t^2\frac{d^2y}{dt^2} - 2t\frac{dy}{dt} + 2y(t) = t^2\left(\frac{d^2z}{dx^2}\cdot\frac{1}{t^2} - \frac{dz}{dx}\cdot\frac{1}{t^2}\right) - 2t\left(\frac{dz}{dx}\cdot\frac{1}{t}\right) + 2z(x)$$

$$= \frac{d^2z}{dx^2} - 3\frac{dz}{dx} + 2z = 0 .$$

Changing the independent variable like this in a DE is done often, because it can reduce a complicated equation to a simpler one.

Now let's use this result to solve an Euler equation that has complex roots.

Example. Solve the Euler DE

$$(E2) \qquad t^2\frac{d^2y}{dt^2} + 5t\frac{dy}{dt} + 8y = 0 .$$

You won't have to do (1) through (5) all over again; just use the substitutions (2) and (5) in the DE (E2), as we did in line (6) above.   If you get stuck, the answers are right after Step 3 below.

1. Use (2) and (5) in equation (E2) and write the constant coefficient DE that has the same eigenvalues. Then find the eigenvalues.

2. Use the eigenvalues to find two linearly independent solutions of the constant coefficient DE you found in Step 1; call these solutions $z_1(x)$ and $z_2(x)$.

3. Now let $x = \ln t$ in your solutions from Step 2 to get two linearly independent solutions $y_1(t)$ and $y_2(t)$ of the DE (E2).

In Steps 1 - 3 above you should have ended up with the constant coefficient DE

$$\frac{d^2z}{dx^2} + 4\frac{dz}{dx} + 8z = 0 \;, \text{ which has eigenvalues } r = -2 \pm 2i \text{ and solutions } z_1(x) = e^{-2x}\cos(2x)$$

and $z_2(x) = e^{-2x}\sin(2x)$. So, solutions of the original Euler DE (E2) are $y_1(t) = t^{-2}\cos(2\ln t)$

and $y_2(t) = t^{-2}\sin(2\ln t)$.

4. Plug these last solutions into the Euler DE (E2) to verify that they are solutions.

Task 1. Solve the following Euler equations by hand using the method of the Example. Then verify your answers by plugging them in, either by hand or using your CAS.

  1. $t^2 y'' - ty' + 2y = 0$
  2. $t^2 y'' - 3ty' + 6y = 0$

Task 2. Use your CAS to solve the DEs of Task 1 directly.

• A first look at systems

Systems of differential equations arise naturally in many areas of study, for example, population dynamics, electrical networks, and mechanics, to name but a few.

Here's an example of a *linear system:*

(1)
$$\frac{dx}{dt} = y$$
$$\frac{dy}{dt} = -x$$

A *solution* of any system is an *ordered pair of functions, (x(t), y(t)) , a vector function*, which satisfies the system on some time interval, $a \le t \le b$. For example, one solution of the system (1) is the vector function

(2)
$$\mathbf{u}(t) = (x(t), y(t)) = (\sin(t), \cos(t)) .$$

(We'll use bold-faced type for solution names to remind you that they're vectors. Also, recall that the vector notation in line (2) means that $x(t) = \sin(t)$ and $y(t) = \cos(t)$.) Let's plug $\mathbf{u}(t)$ into the system (1) to see that it's a solution. We have

$$\frac{dx}{dt} = \frac{d}{dt}(\sin(t)) = \cos(t) = y(t)$$
$$\frac{dy}{dt} = \frac{d}{dt}(\cos(t)) = -\sin(t) = -x(t) .$$

Task 1. Show that $\mathbf{v}(t) = (\cos(t), -\sin(t))$ is also a solution of system (1).

You can see at a glance that neither of the above solutions $\mathbf{u}(t)$ nor $\mathbf{v}(t)$ is a constant multiple of the other (recall that if $\mathbf{u} = (x, y)$ is a vector, then a constant multiple of it would be $k\mathbf{u} = k(x, y) = (kx, ky)$). So we say that $\mathbf{u}(t)$ and $\mathbf{v}(t)$ are *linearly independent vectors* and that they form a *fundamental solution set* for system (1).

Task 2. You might expect, from your earlier study of second order linear equations, that any linear combination $C_1\mathbf{u}(t) + C_2\mathbf{v}(t)$ of solutions of system (1) is also a solution. Show that this is the case. Then, if $\mathbf{u}$ and $\mathbf{v}$ also happen to be linearly independent, then <u>all</u> solutions must look like $C_1\mathbf{u}(t) + C_2\mathbf{v}(t)$, which is called *the general solution.* (See your text for a proof.)

Task 3. Now you can solve some *initial value problems.* Find solutions of system (1) that satisfy each of the following sets of initial conditions. Use the fact that $C_1\mathbf{u}(t) + C_2\mathbf{v}(t)$ is the general solution, and use the initial conditions to set up a pair of algebraic equations to solve for C_1 and C_2. We'll do the first one for you; see the details following this Task.

1. $x(0) = 1$ and $y(0) = 1$.

2. $x(\pi/2) = -2$ and $y(\pi/2) = 0$.

3. $x(\pi) = \sqrt{3}$ and $y(\pi) = 0$.

The general solution of system (1) can be written, using vector algebra, as

$$
\begin{aligned}
(x(t),\, y(t)) &= C_1\mathbf{u}(t) + C_2\mathbf{v}(t) \\
&= C_1\,(\sin(t),\, \cos(t)) + C_2\,(\cos(t),\, -\sin(t)) \\
&= (\, C_1\sin(t) + C_2\cos(t),\ C_1\cos(t) - C_2\sin(t)\,)\,.
\end{aligned}
$$

Thus, we have, by equating components,

(3) $\qquad x(t) = C_1\sin(t) + C_2\cos(t) \quad \text{and} \quad y(t) = C_1\cos(t) - C_2\sin(t)\,.$

Now use the initial conditions $x(0) = 1$ and $y(0) = 1$ in equations (3) to get

(4) $\qquad \begin{aligned} x(0) &= C_1{\cdot}0 + C_2{\cdot}1 = 1 \\ y(0) &= C_1{\cdot}1 - C_2{\cdot}0 = 1\,. \end{aligned}$

Equations (4) have the unique solution $C_1 = C_2 = 1$, so the unique solution of the IVP is

$$(x(t),\, y(t)) = (\,\sin(t) + \cos(t),\ \cos(t) - \sin(t)\,)\,.$$

Task 4.

1. Show that the vector functions $\mathbf{r}(t) = (e^t,\, e^t)$ and $\mathbf{s}(t) = (e^{2t},\, 3e^{2t})$ are solutions of the linear system

$$
\begin{aligned}
\frac{dx}{dt} &= \frac{1}{2}x + \frac{1}{2}y \\[4pt]
\frac{dy}{dt} &= -\frac{3}{2}x + \frac{5}{2}y
\end{aligned}
$$

2. Are $\mathbf{r}$ and $\mathbf{s}$ linearly independent? What is the general solution of the system?

3. Find the solution that satisfies the initial conditions $x(0) = 1$ and $y(0) = 5$.

__Task 5.__ Show that $\mathbf{u}(t) = (1, 0, 0)$, $\mathbf{v}(t) = (t, 1, 1)$, and $\mathbf{w}(t) = (1, t, t^2)$ are solutions of the three-dimensional system

$$\frac{dx}{dt} = \frac{t}{t - 1}\, y + \frac{1}{1 - t}\, z$$

$$\frac{dy}{dt} = \frac{1}{t(1 - t)}\, y + \frac{1}{t(t - 1)}\, z$$

$$\frac{dz}{dt} = \frac{2}{1 - t}\, y + \frac{2}{t - 1}\, z$$

This is an example of a ___linear system with variable coefficients___.

     As you might expect, most systems of DEs cannot be solved in closed form. We must then rely on qualitative analysis, numerical solutions, and pictures to predict how the system will behave. A notable exception is the class of ___first order linear homogeneous systems with constant coefficients___, those of the form

$$\frac{dx}{dt} = ax + by$$

$$\frac{dy}{dt} = cx + dy$$

where $a,\ b,\ c,$ and d are constants. System (1) is an example. We know how to find complete solutions for such systems, and you'll learn how to do this soon.

- **The phase plane**

A picture of the solutions of a system would be enormously helpful in understanding the qualitative behavior of a system. Of course, we can always graph the x-component of a solution on a (t,x)-coordinate system, and similarly graph the y-component, and we often do this. But those graphs don't give us an idea of the interaction between x and y . So we'll graph them both together in the (x,y)-plane, which we'll now call *the phase plane.*

You graphed <u>parametric equations</u> in your calculus course; what we'll do here is the same thing. Think of t as time and think of a solution vector (x(t), y(t)) as a point which moves through the (x,y)-plane as time increases.

Consider the example from the last lesson, the system

(1)
$$\frac{dx}{dt} = y$$
$$\frac{dy}{dt} = -x$$

which has a solution $\mathbf{u}(t) = (\sin(t), \cos(t))$. We want to parametrically graph the equations

(2)
$$x = \sin(t)$$
$$y = \cos(t)$$

in the (x,y)-plane (there is no t-axis).

Task 1. Make a table of values for x and y of equations (2) for $0 \le t \le 2\pi$ and plot the curve (by hand) parametrically in the (x,y)-plane. Indicate the direction of increasing time by drawing little arrows on the curve.

Sometimes (but not often) it's possible to "eliminate the parameter t " and find an equation for the solution curve of a system in terms of x and y alone; then, with any luck, you can recognize the equation as a familiar curve from calculus. In our example, we cleverly notice that

$$x^2 + y^2 = \sin^2(t) + \cos^2(t) = 1 \, ,$$

which is an equation for the the unit circle, so the solution $\mathbf{u}(t)$ traces out the unit circle and moves in a clockwise direction. The circle $x^2 + y^2 = 1$ and its clockwise motion is called a *trajectory* for the system (1). Other terms for it are *path* or *orbit*.

Another solution of system (1) we looked at in the last lesson, namely
v(t) = (cos(t), -sin(t)), traces out the <u>same</u> circle $x^2 + y^2 = 1$. In fact, there are infinitely many solutions that trace out this circle (can you find another one?). So don't confuse a <u>trajectory</u> with a <u>solution</u> - they are not quite the same thing! We'll investigate this further later.

Keep in mind that infinitely many solutions can share the same trajectory.

Task 2.

1. In Task 3 of Lesson 40, you found three different solutions of system (1) that satisfied three different initial conditions. These solutions are:

$$(\sin(t) + \cos(t), \cos(t) - \sin(t)) \quad \text{through the point } (1,1) \text{ at time } t = 0;$$

$$(-2\sin(t), -2\cos(t)) \quad \text{through the point } (-2,0) \text{ at time } t = \pi/2;$$

$$(-\sqrt{3}\cos(t), \sqrt{3}\sin(t)) \quad \text{through the point } (\sqrt{3}, 0) \text{ at time } t = \pi.$$

   What are equations of the the trajectories for these three solutions? (Hint: Compute $x^2 + y^2$ for each one.)

2. Graph these trajectories and indicate on them the direction of increasing t using small arrows.

Task 3. There is one special "curve" representing a solution of system (1) which does not change with time. What is it? (Hint: If x does not change, then dx/dt ≡ 0; same for y.)

You should now have a fairly good picture of the ***phase portrait*** for system (1), that is, a picture of how various solutions move together in the (x,y)-plane. Your picture should look like:

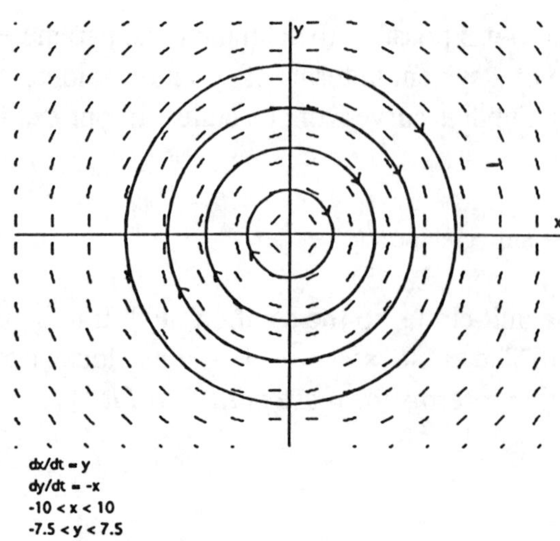

The origin $(0,0)$ is the special "curve" we asked about in Task 3. It corresponds to the ***stationary solution*** $x(t) \equiv 0$, $y(t) \equiv 0$. Such a stationary point is called an ***equilibrium point*** or ***critical point*** for a system - there is no motion there: a solution which begins there will stay there for all time; it's a constant vector function.

Task 4. Here's an interesting and important property that some systems enjoy. Let $\mathbf{u}(t) = (\sin(t), \cos(t))$ be our old solution of system (1).

1. Show that the ***time-translates*** defined by

$$\mathbf{u}(t + c) = (\sin(t + c), \cos(t + c)), \quad \text{where } c \text{ is any constant,}$$

   are also solutions of system (1). (Hint: Plug it in.)

2. What trajectory do the time-translates $\mathbf{u}(t + c)$ follow? (Hint: Again, look at $x^2 + y^2$.)

3. Consider the time-translate solution $\mathbf{u}(t + \pi/4)$. Does it "preceed" the solution $\mathbf{u}(t)$ as they both go around the unit circle, or does it "lag behind"? Where is each solution at time $t = 0$? In which direction are they moving?

- **Three linear systems: Slope fields and phase portraits**

Consider again the system from previous lessons:

(1)
$$\frac{dx}{dt} = y$$
$$\frac{dy}{dt} = -x$$

Let's look at its **slope field** or **direction field** (you did this in Lesson 5 for single first order equations.) We can find the slope of any trajectory for system (1) by using the chain rule:

$$slope\ at\ (x,y)\ =\ \frac{dy}{dx}\ =\ \frac{dy/dt}{dx/dt}$$

and substituting in the values for dy/dt and dx/dt from the system (1) to get:

$$\frac{dy}{dx}\ =\ \frac{-x}{y}\ .$$

This equation, **the slope equation**, is a first order DE, and you can use your CAS to draw its slope field, just as you did in Lesson 5. Do it. The slope field is a picture of the slopes of the trajectories in the phase plane.

In this example, the slope equation is separable, so it can be solved for the trajectory curves. Do so. You should get: $x^2 + y^2 = C$, where C is a non-negative constant. This family is a batch of circles centered at the origin, as you know.

You can then use the system equations (1) to determine the direction of motion of the solution point $(x(t), y(t))$, because dx/dt tells you the point's velocity in the x-direction, and dy/dt tells you its velocity in the y-direction. For example, at the point $(1,1)$, we have dx/dt = y = 1 , which is positive, and dy/dt = -x = -1 , which is negative; so the solution is moving to the right and down. The net result is clockwise motion along the circle $x^2 + y^2 = 1$. Check the signs for some of the other circles.

Task 1. Consider the following system:

(2)
$$\frac{dx}{dt} = x$$
$$\frac{dy}{dt} = 2y$$

1. What is the equilibrium solution?

2. Find the slope equation, $dy/dx = ?$, for this system and use your CAS to draw the slope field in the region $-2 \le x \le 2$, $-2 \le y \le 2$. (See the Appendix for Lesson 5.)

3. Solve the slope equation for the family of trajectories and sketch a few of these trajectories by hand on your slope field graph. Indicate the direction of motion of the solutions along the trajectories. Be sure to include at least one trajectory in each quadrant of the plane. You should get a phase portrait like:

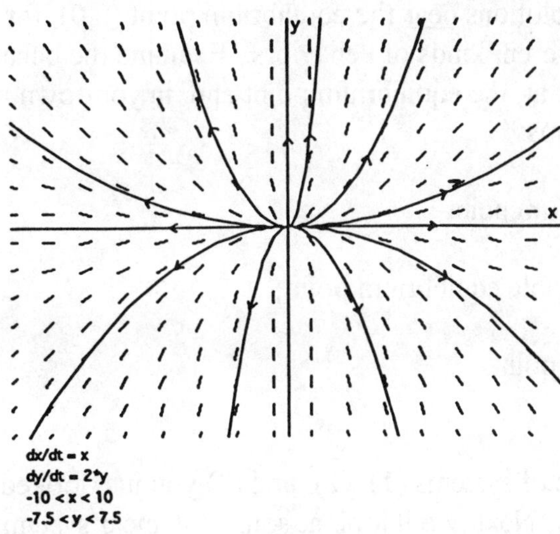

4. Describe the motion of the solution along the trajectory which begins at the equilibrium point. (This is a "trick" question to remind you about constant solutions.)

5. Describe the motion of the solutions along four other trajectories, one in each quadrant. Which way do the solutions move?

Task 2.

1. Show that the vector $\mathbf{u}(t) = (e^t, e^{2t})$ is a solution of system (2).

2. What trajectory does $\mathbf{u}(t)$ follow? Give its (x,y)-equation.
    (Hint: Eliminate the parameter t by using $e^{2t} = (e^t)^2$.)

3. Show that the time-translates $\mathbf{u}(t + c) = (e^{t+c}, e^{2(t+c)})$ are also solutions of (2), where c is any constant. (Hint: Plug in.)

4. What trajectory do the time-translates follow? Give their (x,y)-equation.

5. Does the time-translate $\mathbf{u}(t - 1)$ preceed or lag behind $\mathbf{u}(t)$ as they move along their common trajectory? Where is each of them at $t = 0$? at $t = 1$?

Task 3. Now consider the following system, which is a slight modification of system (2):

$$\frac{dx}{dt} = -x$$

(3)

$$\frac{dy}{dt} = -2y$$

Repeat Steps 1 - 5 of Task 1 for this system. Explain how its phase portrait differs from that of system (2).

Task 4. The motions of solutions <u>near</u> the equilibrium point $(0,0)$ for all three systems (1), (2), and (3) represent three different kinds of behaviors. Examine the behavior of solutions of these three systems <u>near</u>, but not <u>at</u>, the equilibrium point and, in your own words, describe what is meant by the following terms:

    a. an unstable equilibrium point;

    b. an asymptotically stable equilibrium point;

    c. a stable equilibrium point.

       The three examples of systems (1), (2), and (3) you just looked at are all ***linear systems with constant coefficients.*** Next, we'll look at some nonlinear systems and compare behaviors of solutions.

●   **Two nonlinear systems: Slope fields and phase portraits**

Let's look at two ***nonlinear*** systems and see if we can observe some different behaviors from that of linear systems.

Here's a famous example from population dynamics, called a "predator-prey" problem, or, more formally, ***a Lotka-Volterra system***. Let $x(t)$ be the number of prey (bunnies, say) available at time t , and let $y(t)$ be the number of predators (foxes, for example). Then a simple model for the cyclic variations in population sizes is the nonlinear system

$$\frac{dx}{dt} = ax - bxy$$

$$\frac{dy}{dt} = -cy + dxy \,,$$

where a, b, c, d are positive constants which represent birth and death rates of the two species and their rate of interaction. (The xy-terms make this system nonlinear.)   We'll rely on phase portraits and qualitative analysis to see what happens to solutions.

Task 1. Consider the following example of a Lotka-Volterra predator-prey model:

$$\frac{dx}{dt} = 3x - xy$$

(4)

$$\frac{dy}{dt} = -3y + xy$$

1. This time, there are <u>two</u> equilibrium solutions. Find them. (Hint: Set both $dx/dt = 0$ and $dy/dt = 0$ in (4) and solve the equations simultaneously for x and y .)

2. Find the slope equation $dy/dx = ?$. The equation is separable, so solve it. You probably won't be able to identify the trajectory family as any curves you know and it's no wonder - they're the curves $x^3y^3 = Ce^{x-y}$, which are definitely not familiar curves from calculus!

3. Use your CAS to draw the slope field in the region $-5 \le x \le 5$, $-5 \le y \le 5$.

4. Now use your CAS to add some trajectories to your picture, say the ones starting at the points (2,3), (2,5), (-1,1), (-1,-1), (1,-1) whent $t = 0$. (See the Appendix.) Which trajectories are realistic, and which ones would a biologist discard as meaningless? You should get a picture that looks like the one at the top of the next page. We added a few more trajectories.

5. Describe in English the growth and decline of the two species as time passes (the first quadrant behavior of solutions).

6. In this model, is there a set of initial conditions which would result in the extinction of one of the species?

7. Classify the equilibrium points at $(3,3)$ and $(0,0)$ as stable or unstable equilibria. (Remember to look at behavior of solutions <u>close to</u> the critical points.)

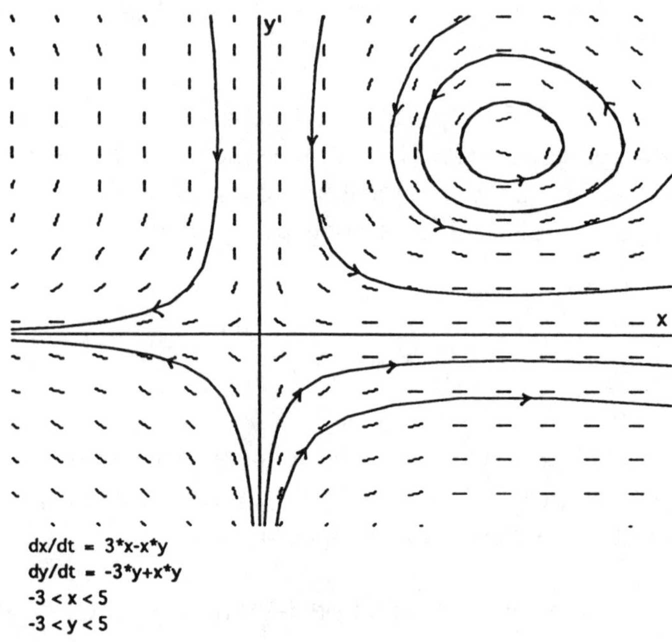

Task 2. A system which arises in the study of vacuum tubes and which has been well-studied is the following **van der Pol system**:

(5)
$$\frac{dx}{dt} = y$$
$$\frac{dy}{dt} = -x + \mu y(1 - x^2) ,$$

where μ is a parameter. Let $\mu = 1$ in the following steps.

1. Find the equilibrium solution(s).

2. The slope field equation for system (5) is

$$\frac{dy}{dx} = \frac{-x + y(1 - x^2)}{y} .$$

Is it linear? separable?

3. We won't ask you to solve the slope field DE, since no one else can either! So we must resort to numerical methods. Use your CAS to draw a slope field and some trajectories in the region $-3 \le x \le 3$ and $-3 \le y \le 3$. Use a stepsize of 0.1 and initial conditions at $t = 0$ of $(0.1, -0.1)$, $(0.5, 0.5)$, $(1,1)$, $(2.5, 2.5)$, $(2.5, -1)$. If you need more trajectories to better see the behavior, add more, or use different initial conditions. Here's a phase portrait:

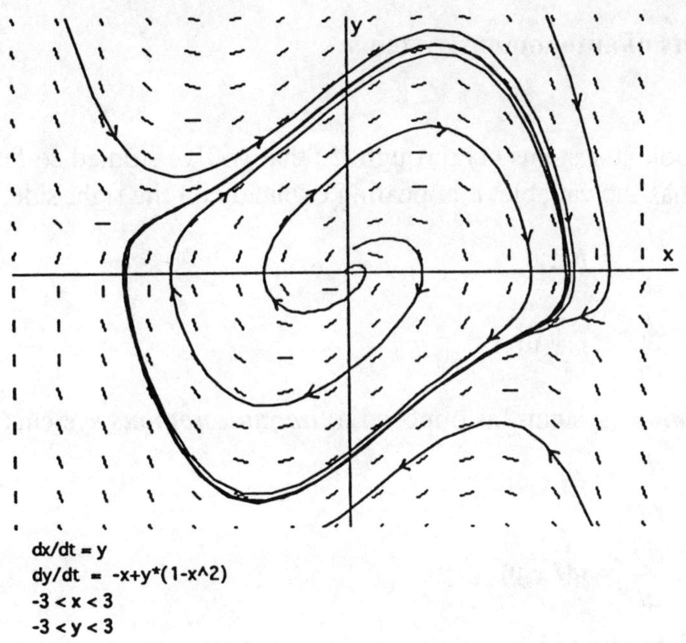

The van der Pol system has one closed trajectory that the other solutions approach asymptotically. Such a closed curve is called a *limit cycle.* Linear systems never have limit cycles. In general, linear systems with constant coefficients exhibit quite simple behavior, which you'll study in some detail. You'll then be able to use your knowledge to predict some behaviors of nonlinear systems.

<u>Task 3.</u> Compare your phase portraits of the three linear systems (1), (2), and (3) that you studied in Lesson 42 with the phase portraits of the above two nonlinear systems (4) and (5). Describe some differences. Describe some similarities.

A note about software: If you have access to either of the DE solvers PHASER for the PC or MacMath for the Macintosh, these computer exercises become much easier. Besides allowing you to watch the trajectories actually being traced out, they also respond more rapidly and are relatively easy to learn, especially MacMath, which is mouse-driven. We used MacMath to generate the phase portraits you see in these Lessons, but we shall continue to write the Appendix guide for Maple and Mathematica. See the Appendix for some software sources and for a list of websites related to DEs.

●   **Phase portraits of autonomous systems**

Take a close look at systems (1) through (5) that you've studied so far in Lessons 42 and 43. Not one of them has the variable t appearing explicitly on the right side; they all look like

(A)
$$\frac{dx}{dt} = f(x,y)$$
$$\frac{dy}{dt} = g(x,y)$$

and are called **autonomous** systems, as opposed to **nonautonomous** systems which have the form

(NA)
$$\frac{dx}{dt} = f(t,x,y)$$
$$\frac{dy}{dt} = g(t,x,y) \ .$$

The three-dimensional system of Lesson 40, Task 5 is nonautonomous, for example.

Autonomous systems have certain properties not shared by most nonautonomous systems, and these properties make their phase portraits manageable. Here are three of the most important ones which you'll investigate:

<u>Property 1</u>. If $\mathbf{u}(t) = (x(t), y(t))$ is a solution of the autonomous system (A), then the **time-translates** $\mathbf{u}(t + c) = (x(t + c), y(t + c))$, where c is any constant, are also solutions of (A).

In Lessons 41 and 42, you showed that this property holds for two autonomous systems. It's possible to prove it for the general autonomous system (A). Try it.

Now, recall from calculus that replacing t by $t + c$ in a function f(t) just shifts the graph of f right or left along the t-axis; <u>the shift does not affect the ouput values of f</u>. For example, if $f(t) = t^2$, the graphs of the time-translates $y = (t + 2)^2$, $y = t^2$, and $y = (t - 1)^2$ are pictured below and <u>all three have the same range of values</u>, namely, $y \geq 0$.

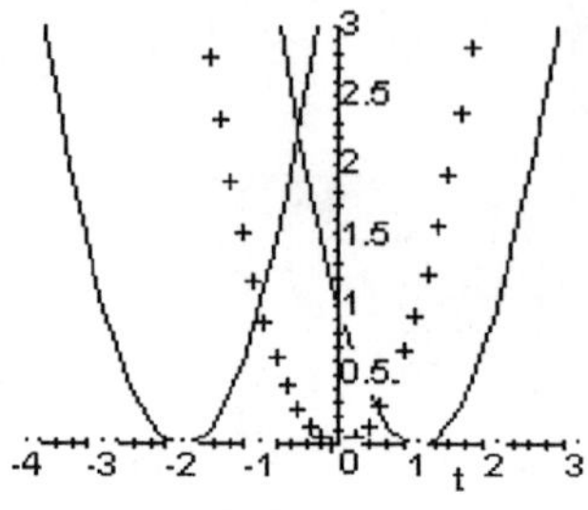

Because the time-translates of a solution are themselves solutions, and all have the same range of values, one effect of Property 1 is:

Property 2. All the time-translates $\mathbf{u}(t + c)$ follow the same trajectory.

Now, the Existence and Uniqueness Theorem (see your text for a formal statement) says that, given a specific time $t = t_0$ and a specific point $P(x_0, y_0)$ in the plane, there is just one solution of a system going through point P at time t_0. This theorem, along with Property 2, allows us to conclude that

Property 3. No two distinct trajectories of an autonomous system (A) can intersect one another.

To convince you that Property 3 holds, suppose that there <u>are</u> two trajectories that intersect at a point P in the plane. Call the trajectories T_1 and T_2. Of course, by the Existence and Uniqueness Theorem, no two solutions can arrive at point P at the same time. So let $\mathbf{u}(t)$ be the solution which follows trajectory T_1 and gets to P at, say, time $t = 2$, that is, $\mathbf{u}(2) = P$; and let $\mathbf{v}(t)$ be the solution which runs along the other trajectory T_2 and gets to P at time $t = 5$, that is, $\mathbf{v}(5) = P$.

By Property 1, the function $\mathbf{z}(t) = \mathbf{u}(t - 3)$, being a time-translate of the solution $\mathbf{u}$, is also a solution of system (A), and Property 2 tells us that $\mathbf{z}(t)$ runs along the trajectory T_1 . Now observe that

$$\mathbf{z}(5) = \mathbf{u}(5 - 3) = \mathbf{u}(2) = P, \text{ and remember that } \mathbf{v}(5) = P ,$$

so that $\mathbf{z}$ and $\mathbf{v}$ are <u>two solutions that go through the same point P at the same time $t = 5$</u>, each following a different trajectory! This contradiction forces us to conclude that T_1 and T_2 can't intersect in the first place.

Look again at your phase portraits of the autonomous systems (1) through (5) of Lessons 42 and 43. Although it might look like some trajectories intersect one another, they never really do. It's an illusion caused by round-off error and/or the thickness of the lines drawn by the computer and/or the fact that a computer graphics screen only contains finitely many grid points on which to draw, so cannot really represent a plane that has uncountably many points in it.

Now you'll verify Properties 1 - 3 using the following simple linear autonomous system

(6)
$$\frac{dx}{dt} = x$$
$$\frac{dy}{dt} = 2y ,$$

which you studied in Lesson 42. Carry out the following steps.

1. System (6) is ***uncoupled***, meaning that dx/dt depends only on x and dy/dt depends only on y. Solve each equation separately and thus show that the general solution of the system is $\mathbf{u}(t) = (C_1 e^t, C_2 e^{2t})$, where C_1 and C_2 can be any constants.

2. From Step 1, if C_1 and C_2 are both positive, then the x and y coordinates of a solution $\mathbf{u}$ are also both positive. This means that the solution is restricted to move only in Quadrant I. Furthermore, we have $dx/dt = C_1 e^t$ and $dy/dt = 2C_2 e^{2t}$, both of which are positive; so the solutions move away from the equilibrium point at the origin. Eliminate the parameter t between $x = C_1 e^t$ and $y = C_2 e^{2t}$ to show that the trajectories are the half-parabolas $y = (C_2/C_1^2)x^2$ in Quadrant I.

3. Where and how do solutions move if $C_1 < 0$ and $C_2 > 0$? Eliminate t and write the (x,y)-equations for the trajectories. Do the same for the other two sign cases. Sketch some of these trajectories along with their arrows.

4. Describe the trajectories if $C_1 = 0$ and $C_2 > 0$; if $C_1 = 0$ and $C_2 < 0$. What of the cases when $C_1 \neq 0$ and $C_2 = 0$? What happens when $C_1 = C_2 = 0$? (Think "equilibrium".)

From your analysis above, you see that a trajectory is either the point at the origin (the equilibrium solution), or a half-parabola, or a half-axis. None of these curves intersect, thus verifying Property 3, that no trajectories intersect - they just appear to intersect at $(0,0)$.

5. To help illustrate Properties 1 and 2, draw the solution of system (6) that passes through the point $(1,1)$ at time $t = 0$, i.e., the solution $\mathbf{z}(t) = (e^t, e^{2t})$ which follows the trajectory $y = x^2$ in the first quadrant. Notice that $\mathbf{z}(t) \to (0,0)$ as $t \to -\infty$ and that $\mathbf{z}(t) \to (+\infty, +\infty)$ as $t \to +\infty$, so that the solution moves from "right next to" the origin at $t = -\infty$ outwards along the half-parabola $y = x^2$ as $t \to +\infty$.

6. Next, you'll look at the time-translates of the solution $\mathbf{z}(t)$ defined in Step 5, namely, the functions

$$\mathbf{w}(t) = \mathbf{z}(t + c) = (e^{t + c}, e^{2(t + c)}), \text{ where } c \text{ is any constant.}$$

Show that the functions $\mathbf{w}(t)$ are solutions of system (6), thus illustrating Property 1.

7. Eliminate the parameter t in the solutions $\mathbf{w}(t)$ of Step 6 and show that these solutions all run along the half-parabola $y = x^2$ in the first quadrant. This illustrates Property 2, that all the time-translates of a solution run along the same trajectory.

8. To illustrate that two solutions can go through the same point at <u>different</u> times, consider again the solution $z(t) = (e^t, e^{2t})$ which goes through the point $(1,1)$ at time $t = 0$. Find a time-translate of z which goes through the point $(1,1)$ at time $t = 2$.

   Properties 1, 2, and 3 will be true for any autonomous system, but <u>not</u> necessarily true for all nonautonomous systems, as you'll demonstrate in the next lesson.

- **Phase portraits of nonautonomous systems**

Now we'll modify system (6) of the last lesson to make it an easy-to-solve nonautonomous system.

$$\frac{dx}{dt} = \frac{1}{t}x$$

(7)

$$\frac{dy}{dt} = y$$

This is a linear system with variable coefficients, and it's nonautonomous. (Don't confuse nonlinear with nonautonomous, or linear with autonomous.)

You're going to show that Properties 1, 2, and 3 of the last lesson do <u>not</u> hold for this system. So do the following steps.

1. Find the general solution of system (7). Since the system is uncoupled, you can solve each equation separately.

2. In Step 1, you should have found the general solution to be $\mathbf{u}(t) = (C_1 t, C_2 e^t)$, for $t \neq 0$. Eliminate the parameter t and thus show that the solutions follow the trajectories $y = C_2 \exp(x/C_1)$.

3. The solutions that have C_1 and C_2 both positive are restricted to move in Quadrant I when $t > 0$, and they move away from the equilibrium point $(0,0)$. Why? Sketch some of these trajectories (by hand). Determine the trajectories, and sketch some, for the other sign cases of C_1 and C_2. Don't forget the cases when either $C_1 = 0$ or $C_2 = 0$.

4. Let's time-translate the general solution $\mathbf{u}$ of Step 2 in the usual way by defining

$$\mathbf{w}(t) = \mathbf{u}(t + c) = (C_1(t + c), C_2 e^{t + c}) .$$

Show that $\mathbf{w}(t)$ is <u>not</u> a solution of system (7) no matter what constant c is (except $c = 0$). Thus, Property 1 doesn't necessarily hold for nonautonomous systems.

5. Now use the general solution $\mathbf{u}$ defined in Step 2 to find the solution that goes through the point $(1,1)$ at time $t = 1$. Then find the one that goes through the same point $(1,1)$ at time $t = 2$; and find another one that goes through $(1,1)$ at time $t = 3$.

6. Take your three solutions from Step 5 and eliminate t in each one to find the (x,y)-equations of their trajectories. You should get the three different curves: $y = e^{x-1}$, $y = e^{2(x-1)}$, and $y = e^{3(x-1)}$, all going through $(1,1)$. Sketch them on the same graph. Thus, Property 3 doesn't hold for this example: different trajectories <u>can</u> intersect.

Now you can see that the phase plane for this system will be cluttered with intersecting trajectories! The picture can get really messy. This is why we don't draw phase portraits for most nonautonomous systems; it's hard to see what's going on.

Notice that the slope equation for this system is $\dfrac{dy}{dx} = \dfrac{ty}{x}$, which changes with time, so the slope field is constantly changing.

Don't conclude that <u>all</u> nonautonomous systems behave this way. For example, the nonautonomous system

(8)
$$\frac{dx}{dt} = \frac{1}{t}x$$
$$\frac{dy}{dt} = \frac{2}{t}y$$

has trajectories that don't intersect, as you'll now show.

Task 1.

1. Find the general solution of system (8) - it's uncoupled.

2. Eliminate the parameter t in the general solution and find an (x,y)-equation for the family of trajectories. Don't combine the constants C_1 and C_2 . Do two different trajectories ever intersect?

3. The general solution from Step 1 is $\mathbf{u}(t) = (C_1 t,\ C_2 t^2)$ for $t \neq 0$. Is it true that the time-translates $\mathbf{u}(t + c)$ are also solutions?

4. Find the three solutions that go through the point $(1,1)$ at times $t = 1$, $t = 2$, and $t = 3$. Eliminate t in each solution and find (x,y)-equations for their trajectories. Do the three solutions all run along the same trajectory? Compare this behavior to that of system (7).

5. What is the slope equation for this system? Does it change with time?

● **The spring again: An equivalent system and its phase portrait**

Recall the DE for the motion of a mass on a spring (see Lesson 21):

(spring)
$$m\frac{d^2y}{dt^2} + b\frac{dy}{dt} + ky = 0 \ ,$$

where m is the mass of the object, b is the damping factor, and k is the spring constant. A solution $y(t)$ gives the position of the mass at time t, the equilibrium position is $y = 0$, and positive displacement of the mass is measured upward.

We can easily transform equation *(spring)* into a system of two first order equations as follows:

1. Let $v(t)$ be the velocity of the mass. How are $v(t)$ and $y(t)$ related?

2. Since $v(t) = y'(t)$ in Step 1, we then have $v'(t) = y''(t)$. Use the equation *(spring)* to express $v'(t)$ in terms of $v(t)$ and $y(t)$.

Combining your answers to Steps 1 and 2 gives the equivalent first order system of DEs

(SS)
$$\frac{dy}{dt} = v$$
$$\frac{dv}{dt} = -\frac{k}{m}y - \frac{b}{m}v \ .$$

The second order DE *(spring)* and the first order system (SS) are ***equivalent*** in the following sense: If $y(t)$ is a solution of *(spring)*, then the vector $\mathbf{u}(t) = (y(t), y'(t))$ will be a solution of the system (SS). Conversely, if the vector $(y(t), v(t))$ is a solution of the system (SS), then its first component, $y(t)$, will be a solution of the second order equation *(spring)*.

Since you already know how to solve equation *(spring)* - remember the characteristic equation and eigenvalues? - you know all the solutions of system (SS).

One advantage of transforming *(spring)* into a system is that it allows us to study the behavior in the phase plane. Since $y(t)$ and $v(t)$ for a spring are usually bounded functions, we can see the entire motion of the mass in a bounded region of the phase plane and not have to deal with the infinite extent of a t-axis.

<u>**Task 1.**</u>

1. Assume that the mass and spring constants are each 1 , and that the damping factor is 0 . What is system (SS) now? This system should look familiar to you from Lessons 41 and 42. What are the equilibrium points? What is the general solution? Describe the trajectories.

2. Look at the trajectory that starts at the point (2,0) when $t = 0$. Remember that the horizontal axis now represents the displacement, $y(t)$, of the mass, while the vertical axis represents its velocity, $v(t)$. Choose eight evenly spaced points on the trajectory and describe what the mass is doing at the time the trajectory crosses each point. Your description should include phrases like "moving up" and "moving down" , "accelerating" and "decelerating". Where is the mass when its velocity is zero? What is the velocity when the mass reaches its maximum displacement and when does this happen? What is the displacement when the mass is going fastest?

<u>**Task 2.**</u>

1. Use your CAS to draw phase portraits of system (SS) when $m = k = 1$ and
     a) you choose b to give underdamping;
     b) you choose b to give overdamping;
     c) you choose b to give critical damping.

2. For each of the above cases, identify the equilibrium position and say whether it's stable or unstable. What consequences does the stability property of the equilibrium point have for the motion of the mass?

    The system (SS) is another example of a <u>linear system with constant coefficients,</u> and the phase portraits you observed in parts a), b), and c) of Task 2 represent three more kinds of behavior that can happen in linear systems:

    a) In the case of underdamping, the equilibrium point (0,0) is called a ***spiral sink*** or a *focus*. The solutions spiral in towards (0,0) as t gets large, and the smaller the damping factor, the more turns there are in the spiral. This picture makes sense, since the mass is bobbing up and down on the spring, but the amplitude is getting smaller. Thus, both y and v keep changing sign but are getting smaller and smaller as time goes on. Here's a picture:

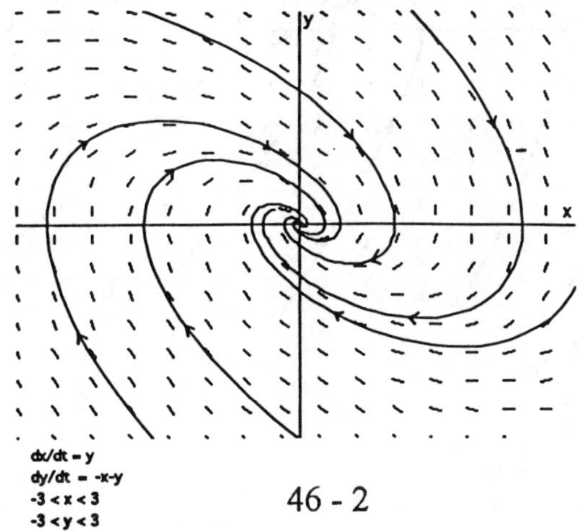

b) In the case of overdamping, the equilibrium point $(0,0)$ is called a ***node***. Most solutions appear to be tangent to a certain line as they approach $(0,0)$. But there's another batch of solutions that march straight into the origin along one of two halves of another line. So there are four half-line trajectories going toward $(0,0)$. The rest of the trajectories (except the origin) are curved. A portrait is:

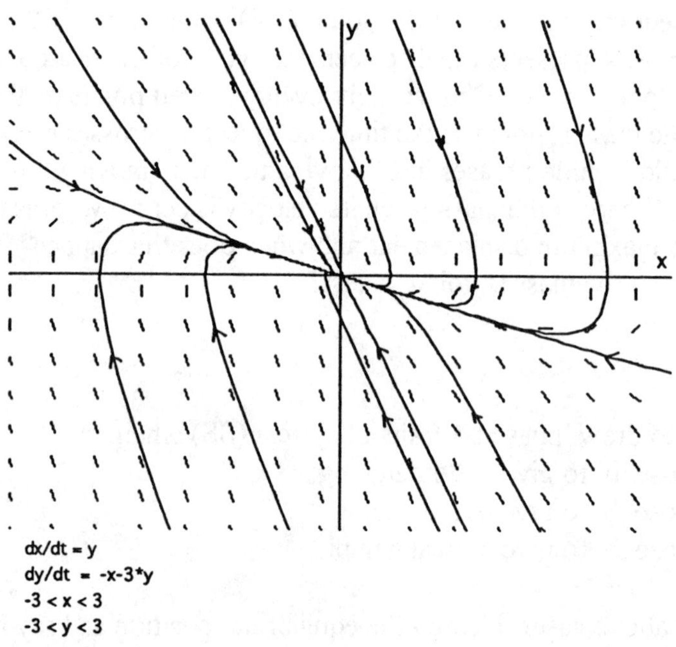

c) When the system is critically damped, the equilibrium point $(0,0)$ is again called a ***node***, sometimes referred to as an ***improper node***. The phase portrait looks much like the picture in case b), but this time there is one line through the origin that all other nontrivial solutions are tangent to. That line is actually two half-line trajectories going toward $(0,0)$. Here's a picture:

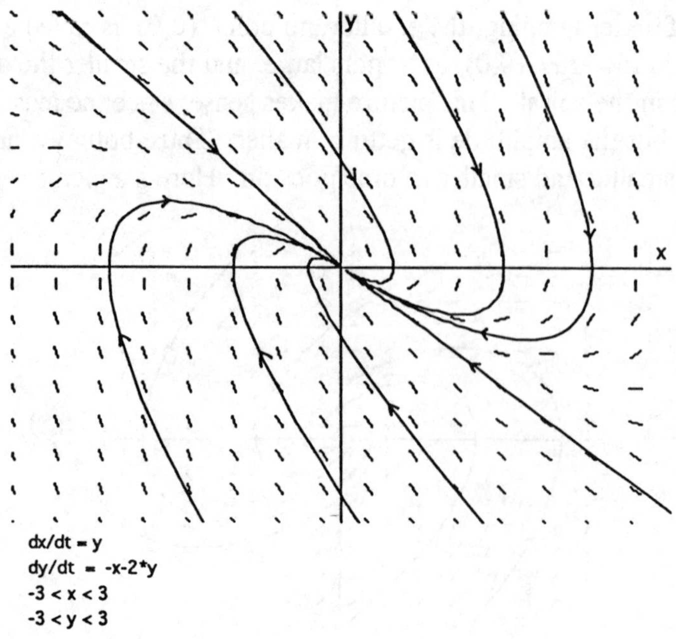

In all three cases of the damped mass-spring behavior, the origin is an asymptotically stable critical point, because any solution that starts "near" $(0,0)$ ends up approaching $(0,0)$ as $t \to +\infty$.

For the undamped spring, when $b = 0$, the trajectories are circles or ellipses centered at the origin. In this case the equilibrium point at the origin is called a ***center***, and it's stable, because any solution that starts "near" $(0,0)$ stays near $(0,0)$ as $t \to +\infty$.

Every second order DE of the form

(1) $$x'' = F(t, x, x')$$

can be written as an equivalent system of two first order DEs. Just introduce a new variable y as the derivative of x,

$$x' = y,$$

and use it in equation (1) by writing $x'' = y' = F(t, x, y)$. So the equivalent system is

$$\frac{dx}{dt} = y$$

$$\frac{dy}{dt} = F(t, x, y) \ .$$

You can do this for higher order equations, too.

Task 3. Write equivalent first order systems for each DE below.

1   A forced harmonic oscillator equation: $my'' + by' + ky = \sin(t)$.

2.   An Euler equation: $t^2 x'' + 2tx' + 3x = 0$.

3.   A third order equation: $x''' - 3x'' + 2x' + 5x = 0$. (Hint: Introduce two new variables y and z as follows: $x' = y$ and $y' = z$. Then what is z' ?)

• **Derivation of the simple pendulum equation**

Let's look at the motion of a simple pendulum, which is related to the motion of a mass on a spring. Suppose we have a bob of mass m attached to a rigid pendulum of length L (see the figure). The bob is pulled up to an initial angle of θ_0 and allowed to fall under its own weight. We'll assume a frictionless pendulum and no air resistance, to keep things simple. Also, assume that the mass of the pendulum arm is negligible compared to the mass of the bob.

Pendulum

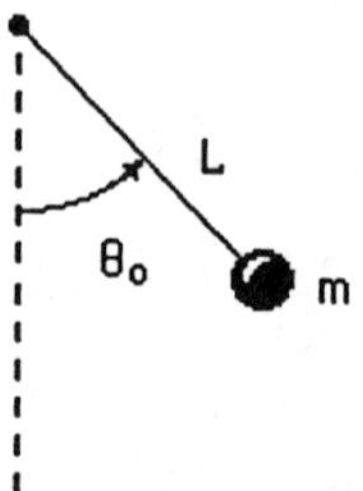

You're going to derive the differential equation of motion of the bob, using Newton's Second Law (again!) and a little geometry.

1. The position of the bob at time t is its location along a circular arc. Draw in this arc, and label it as an axis; call it the s-axis. Where is the best place for you to put the origin? Now choose a positive direction along the arc. Newton's Second Law will require the use of this axis, which measures displacement of the bob in linear units (feet, meters, whatever).

2. The variable most useful in describing the position of the bob at any time t is the angle θ (in radians) which the pendulum makes with the vertical. Label this angle in the picture. In which direction is it measured positively?

3. So that we're using the same frame of reference, let's agree to put the origin at the point where the bob would hang motionless, and take the positive direction to be upward to the right, in other words, counterclockwise along the circle. Then angle θ is also measured positively in the counterclockwise direction. What is the relationship between s and θ? (Do you recall this from calculus? It's $s = L\theta$, where θ is measured in radians, not degrees.)

4. What forces are acting on the mass at any given time? Draw in these force <u>vectors</u>. Remember, there's no damping, air resistance, or friction.

5. The bob will move along a circular arc, starting from the initial position θ_0. Draw a <u>vector</u> to show this initial direction of motion. Do you have enough forces to account for the pull in this direction? (Remember the parallelogram law of addition for vectors. Your resultant force should be pointing in the direction of motion.)

6. If you don't have enough forces to account for the motion you predict, you may be puzzled; the mystery force is the <u>tension</u> on the pendulum, which always pulls the mass back toward the point where the pendulum is attached at the top. Draw in this vector along the pendulum arm. Now do your force vectors add up to a net force in the direction of motion?

7. Now you want to compute the magnitude and direction of the <u>net force</u> (vector sum) acting on the bob in terms of the variable θ and the constants m and g (acceleration due to gravity). To accomplish this, break up the gravitational force vector into two components: one that is tangent to the circle and pulls the mass downward, and the other that is equal in size to the tension but opposite in direction. See the figure. (If this second component didn't balance the tension, the bob would either fly off the pendulum or be pulled up toward the pivot point.)

Pendulum 2

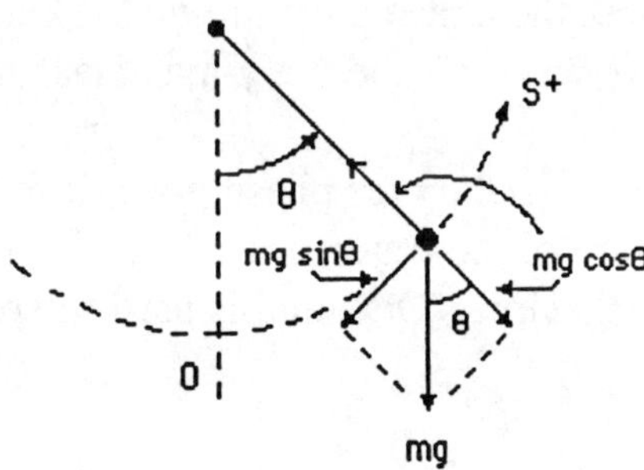

8. Write Newton's Second Law as it applies to this system, using θ as the dependent variable. (Hint: The acceleration is <u>not</u> θ'', it's s''. How are θ'' and s'' related?)

9. Now, using your answer to Step 8, write a DE in terms of θ and the constants alone.

10. What order is this DE? Is it linear? What methods can be used to investigate its behavior? What is (are) the initial condition(s)?

You should end up with the DE $mL\dfrac{d^2\theta}{dt^2} + mg\sin(\theta) = 0$, which simplifies to

$\theta'' + \dfrac{g}{L}\sin(\theta) = 0$, and the initial conditions are $\theta(0) = \theta_0$ and $\theta'(0) = 0$. It's a second

order <u>nonlinear</u> equation (because of the $\sin(\theta)$ term).

- **Simple pendulum behavior**

Now you'll look at the pendulum's phase portrait and describe its behavior.

1. Write the DE for the simple pendulum from Lesson 47, and convert it into a system in the variables x and y, letting $x = \theta$ and $y = d\theta/dt$.

2. You should now have the system

(P)
$$\frac{dx}{dt} = y$$
$$\frac{dy}{dt} = -\frac{g}{L}\sin(x).$$

How many equilibrium points does it have, and where are they?

3. Draw, by hand, what you think the phase portrait will look like. Use $L = g$ for simplicity. Remember that the x-axis measures θ and the y-axis measures $d\theta/dt$.

4. Use your CAS or a DE-solver to get a nice phase portrait for the system (use $L = g$), and indicate the direction of motion of the trajectories. Besides the equilibrium points, you should see three different kinds of behavior. (One kind is hard to spot.) Your picture should look like this:

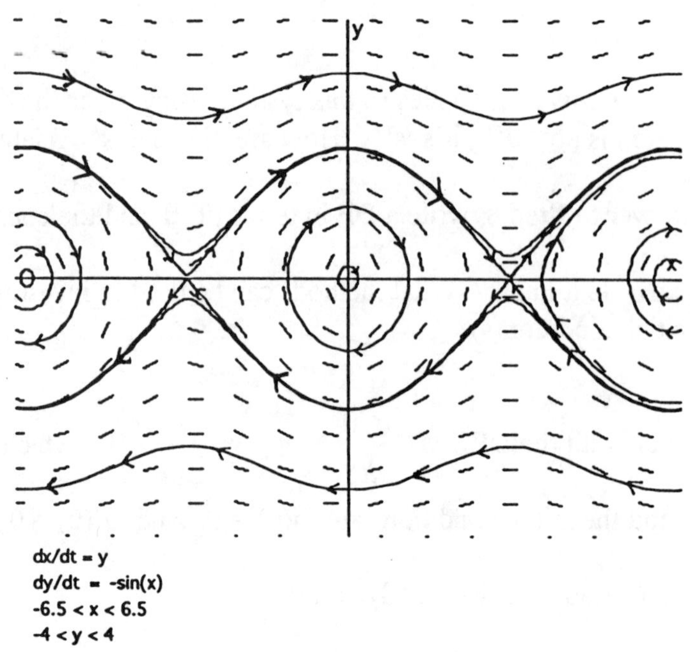

5. Look at the equilibrium points $(n\pi, 0)$, where n is an <u>even</u> integer, that is, $(0,0)$, $(2\pi,0)$, $(-2\pi, 0)$, etc. What does the pendulum do if you start it off at one of these points? Do you think these are stable or unstable equilibria?

6. Now look at the equilibrium points $(n\pi, 0)$, where n is an <u>odd</u> integer. What is the pendulum's motion when you start it off here?

7. The answer to Step 6 is: "It stays there!" Hard to believe, but the pendulum is perfectly balanced in an upside down position! Do you think the points $(n\pi, 0)$, where n is an odd integer, are stable or unstable equilibria?

8. Look at one of the trajectories that is a closed curve. Explain the motion of the pendulum corresponding to this trajectory; for example, what is its maximum θ-displacement? Where is the velocity zero, and what's happening to the pendulum then? Where is the pendulum bob when the velocity is a maximum? Is this realistic behavior? These closed trajectories encircle the equilibrium points $(n\pi, 0)$, where n is an <u>even</u> integer; so these are stable equilibrium points.

9. Now look at one of the trajectories that is a wavy curve. Explain the motion of the pendulum corresponding to this trajectory. In what direction does the trajectory flow? What does this tell you about the motion of the pendulum? There are some identical trajectories elsewhere in the plane, but the flow is in the opposite direction. What do they say about the motion?

10. There is one last kind of trajectory, the hard-to-spot kind. To find one, look at $(-\pi,0)$. Start a solution very near it, a little above and to the right. What happens to this trajectory? If you move the pendulum bob into the almost vertical position $(\theta \approx -\pi)$ and give it just the littlest bit of initial velocity $(\theta' \approx 0.1)$, it will swing around just one full revolution and slow down, slow down, slow down as it approaches the $(\pi,0)$ position. Remember that trajectories don't intersect, so it never quite reaches $(\pi,0)$. Not very realistic behavior, but this is, after all, an ideal pendulum. This trajectory is called a ***separatrix***, because it separates two different kinds of behaviors. When n is an odd integer, the equilibrium points $(n\pi,0)$ are unstable equilibria like this.

11. Now solve the slope equation $\dfrac{dy}{dx} = -\dfrac{\sin(x)}{y}$ for the trajectory curves. See if you can determine what values of the integration constant C give the closed trajectories, what values of C give the wavy trajectories, and what values of C give the separatrices. (Hint: You should get the equations $y = \pm\sqrt{2 - \cos(x)}$ for the separatrices.)

●　　　**Eigenvalues and eigenvectors**

Now you'll learn how to find the solutions of a *linear, homogeneous, constant coefficient system.* It's a system of the form

(1)
$$\frac{dx}{dt} = ax + by$$
$$\frac{dy}{dt} = cx + dy$$

where $a,\ b,\ c,\ d$ are constants. The term <u>homogeneous</u> is used because the constant vector $(x(t),\ y(t)) = (0,0)$ is a solution, an equilibrium solution. Another way to say this is that there are no forcing functions $f(t)$ and $g(t)$ on the right-hand side of equations (1).

Let's find all the equilibrium points of system (1). That means we must solve the algebraic system

(2)
$$ax + by = 0$$
$$cx + dy = 0 \ .$$

Notice that each equation in system (2) represents a line through (0,0). Either one of two things can happen:

　　　a)　　　the lines are the same; or
　　　b)　　　the lines are distinct.

If the lines are the same, then their coefficients must be proportional, so $a/c = b/d$, or $ad - bc = 0$. In this case, every point on the single line $ax + by = 0$ is a solution of system (2), so each one is an equilibrium point of system (1).

If the lines are distinct, then $ad - bc \neq 0$, and the origin is the only point of intersection - the only equilibrium point. We say that system (1) has an *isolated equilibrium point* at (0,0). This is the important case and the one we'll consider now.

To motivate our work, recall the second order mass-spring DE

(*spring*)　　　　　　　　$$my'' + by' + ky = 0$$

and its equivalent system

$$\frac{dy}{dt} = v$$

(SS)

$$\frac{dv}{dt} = -\frac{k}{m}y - \frac{b}{m}v$$

from Lesson 46.

Since (spring) has exponential functions for solutions, so does system (SS), and system (SS) is an example of system (1); just let $a = 0$, $b = 1$, $c = -k/m$, $d = -b/m$ in system (1) and you'll get system (SS). So, to solve systems like (1), it makes sense to try exponential functions as solutions.

Example. Here's a nice system:

(3)

$$\frac{dx}{dt} = x + y$$

$$\frac{dy}{dt} = 3x - y \ .$$

We'll try the vector solution

$$\mathbf{u}(t) = (Ae^{rt}, Be^{rt}) \ ,$$

where A, B, and r are constants to be found. Plug this vector into system (3) to get

$$Are^{rt} = Ae^{rt} + Be^{rt}$$

$$Bre^{rt} = 3Ae^{rt} - Be^{rt} \ .$$

Since e^{rt} is never zero, we can divide each of the above equations by it and rearrange terms to get (do the algebra):

(4)

$$0 = A(1 - r) + B$$

$$0 = 3A - B(1 + r) \ .$$

System (4) is two equations in the three unknowns A, B, and r, so there won't be a unique solution. In an (A,B)-plane, the equations (4) are lots of different pairs of lines through the origin $(A,B) = (0,0)$. The solution $(A,B) = (0,0)$ would give us $\mathbf{u}(t) = (0,0)$, which isn't very informative. So what we'll do is make the lines in (4) identical; then we can choose any point (A,B) on the single line to fill out our solution $\mathbf{u}(t)$. To make the lines identical, their slopes must be equal, so the coefficients of A and B must be proportional. Thus we have

$$\frac{1 - r}{3} = \frac{1}{-(1 + r)} \ ,$$

which simplifies to (do the algebra)

(5)

$$r^2 - 4 = 0 \ .$$

Equation (5) is called *the characteristic equation* for system (3). It has two roots, $r_1 = 2$ and $r_2 = -2$, called *eigenvalues*. So we now have two possible families of solutions:

$$\mathbf{u}(t) = (Ae^{2t}, Be^{2t}) \quad \text{and} \quad \mathbf{v}(t) = (Ce^{-2t}, De^{-2t}).$$

(The same constants A and B won't work in **v** as work for **u**, so we changed their names to C and D.)

Remember that we deliberately made these two choices of $r = \pm 2$ to make the lines in system (4) identical, so we can find A, B, C, and D by using system (4). First, substitute $r_1 = 2$ into system (4) to get the equations

(6)
$$0 = -A + B$$
$$0 = 3A - 3B.$$

These lines are identical (we planned it that way) and tell us that we must have A = B. These infinitely many solution points (A,A) are called *eigenvectors* associated with the eigenvalue $r_1 = 2$. We finally have some of our desired solutions for system (3), namely,

$$\mathbf{u}(t) = (Ae^{2t}, Ae^{2t}), \quad \text{where A is any constant.}$$

Now we'll do the same thing with the other eigenvalue, $r_2 = -2$, to get a different family of solutions. Substitute $r_2 = -2$ into equations (4) and don't forget to change A and B to C and D. We get the equations (check this)

$$0 = 3C + D$$
$$0 = 3C + D,$$

so any pair (C,D) with D = -3C will work. The resulting solutions (C, -3C) are the eigenvectors corresponding to the eigenvalue $r_2 = -2$, and so another set of solutions for system (3) is

$$\mathbf{v}(t) = (Ce^{-2t}, -3Ce^{-2t}), \quad \text{where C is any constant.}$$

We're only going to need one solution from each of the families $\mathbf{u}(t)$ and $\mathbf{v}(t)$, so we can pick values of A and C that will give us "nice" answers. Here, A = 1 and C = 1 will do. We finally end up with the two different solutions

$$\mathbf{u}(t) = (e^{2t}, e^{2t}) \quad \text{and} \quad \mathbf{v}(t) = (e^{-2t}, -3e^{-2t}).$$

These solutions are linearly independent, since neither vector is a constant times the other. The *general solution* of system (3) is, finally, a linear combination of **u** and **v**, that is,

(7) $C_1\mathbf{u}(t) + C_2\mathbf{v}(t) = (C_1e^{2t} + C_2e^{-2t}, C_1e^{2t} - 3C_2e^{-2t})$, with C_1 and C_2 any constants.

Don't make the mistake of "absorbing" the -3 into the C_2 and calling the result C_3 in the general solution (7); you'll lose pertinent information about the solutions if you do that.

<u>**Task 1.**</u>

1. Use your CAS to solve system (3). Does the computer solution agree with the general solution we derived?

2. Solve the **_initial value problem_**: Find the solution of system (3) that goes through the point (1,2) at time $t = 0$. Check your answer by plugging it into the system.

3. Now use your CAS to solve the IVP of Question 2.

<u>**Task 2.**</u> In this task you can study the phase portrait of system (3).

1. Use your CAS to get a nice picture of the phase portrait in the region $-3 \le x \le 3$ and $-3 \le y \le 3$. Is the origin a stable or unstable equilibrium point? It's called **_a saddle point_**. You'll get this kind of picture whenever the eigenvalues are real, nonzero, and of opposite signs. Here's a portrait:

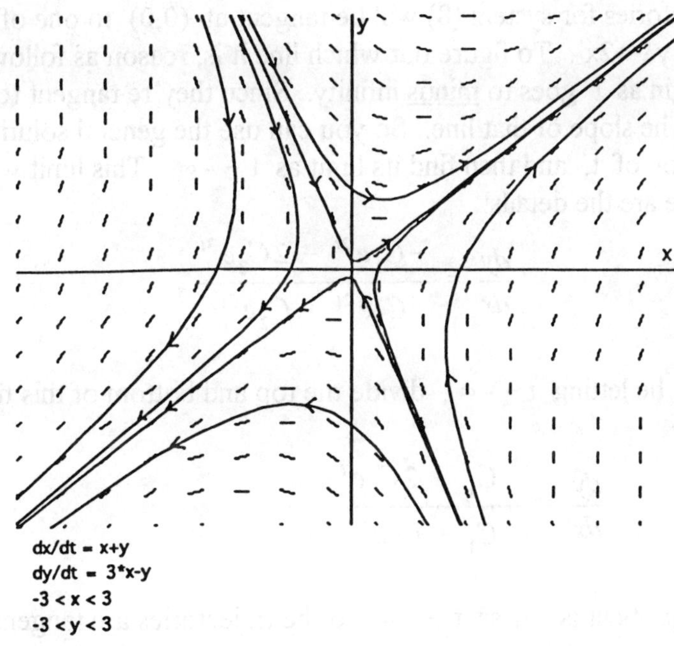

2. Besides the origin, you should see four half-line trajectories in your picture, and the rest curves. To see where the half-line trajectories came from, let $C_1 = 0$ in the general solution (7). You get the solution curves $w(t) = (C_2 e^{-2t}, -3C_2 e^{-2t})$, which run <u>toward</u> the origin along the two half-lines that make up the line $y = -3x$ (eliminate the parameter t). This is the line of eigenvectors for $r_2 = -2$. Now let $C_2 = 0$ in the general solution. You should get a family of solutions which run <u>away</u> from the origin along the two half-lines that make up the line $y = x$, the other line of eigenvectors.

<u>Task 3.</u>

1. Use the method of this lesson to find the general solution of the system

(8)
$$\frac{dx}{dt} = x - y$$
$$\frac{dy}{dt} = 2x + 4y .$$

Your eigenvalues should be $r_1 = 2$ and $r_2 = 3$.

2. Now use your CAS to find the general solution of this system.

<u>Task 4.</u> Get a nice phase portrait of system (8). The equilibrium point $(0,0)$ is a **node**, and it's unstable. All the trajectories run away from the origin. Again, you should have four half-line trajectories corresponding to the lines $y = -x$ and $y = -2x$, which are the lines of eigenvectors. There's a phase portrait of a node in Lesson 46 - the overdamped spring.

The trajectories for system (8) will be tangent at $(0,0)$ to one of the lines of eigenvectors, either $y = -x$ or $y = -2x$. To figure out which line it is, reason as follows: The trajectories approach the origin as t goes to <u>minus</u> infinity. Since they're tangent to a line there, their slopes should approach the slope of that line. So you can use the general solution to find the slope dy/dx as a function of t, and then find its limit as $t \to -\infty$. This limit will be the slope of the desired line. Here are the details:

The slope is
$$\frac{dy}{dx} = \frac{-C_1 e^{2t} - 2C_2 e^{3t}}{C_1 e^{2t} + C_2 e^{3t}} .$$

Now, since you'll be letting $t \to -\infty$, divide the top and bottom of this fraction by e^{2t} to get

$$\frac{dy}{dx} = \frac{-C_1 - 2C_2 e^{t}}{C_1 + C_2 e^{t}} .$$

The limit of this quotient is -1 as $t \to -\infty$, so the trajectories are tangent to the line of eigenvectors $y = -x$.

● **Repeated eigenvalues**

We classify the isolated equilibrium point (0,0) of the linear system

(L)
$$\frac{dx}{dt} = ax + by$$
$$\frac{dy}{dt} = cx + dy$$

in one of five different ways, depending on the values of the eigenvalues (the characteristic equation will be quadratic). These classifications are:

EIGENVALUES	EQUILIBRIUM POINT (0,0)
1. real, unequal, opposite sign	*saddle*
2. real, unequal, same sign	*proper node*
3. real, equal, non-zero	*improper node*
4. complex, real part nonzero	*focus*, or *spiral*
5. complex, real part zero	*center*

The case when one or both of the eigenvalues is zero will be handled later.

So far in these Lessons, you've seen phase portraits of all five of these types, but you've only learned to solve Types 1 and 2, in Lesson 49. Now we'll work on solving the other three types. In the following example that you're going to work through, if you get stuck on a step, go on to the next one and you'll find the answer there.

Example. Consider the system

(1)
$$\frac{dx}{dt} = 3x - y$$
$$\frac{dy}{dt} = x + y \, .$$

1. Use the method of Lesson 49 to find the characteristic equation and eigenvalues for this system. (There's a "short-cut" method for doing this if you know about matrices. Consult your text.)

2. In Step 1 you should have $r^2 - 4r + 4 = 0$ with the repeated root $r = 2$. Find the eigenvectors for $r = 2$ and write the corresponding family of solutions of system (1).

3. The result of Step 2 is the family of solutions $\mathbf{u}(t) = (Ae^{2t}, Ae^{2t})$ and you can use $A = 1$ to get a single basic solution $\mathbf{u}(t) = (e^{2t}, e^{2t})$. How can we find another, linearly independent, solution? In the case of a second order equation in Lesson 23, you multiplied your first solution by t to get a second solution. Show that this doesn't work here; i.e., show that the vector $\mathbf{z}(t) = (Ate^{2t}, Bte^{2t})$ is <u>not</u> a solution of system (1) unless $A = B = 0$.

4. What else might you try? Do you remember the <u>reduction of order method</u> from Lesson 38 ? With that method, you try for another solution of the form $(g(t)e^{2t}, h(t)e^{2t})$, where $g(t)$ and $h(t)$ are functions to be found. We'll save you some work: use linear functions $g(t) = At + B$ and $h(t) = Ct + D$, and find the constants A, B, C, and D so that the vector

$$\mathbf{v}(t) = ((At + B)e^{2t}, (Ct + D)e^{2t})$$

is a solution. Go ahead, plug it into system (1). What set of equations do you get?

5. At this point, you should have the system

$$(2) \qquad \begin{aligned} (-A + C)t + (-B + A + D) &= 0 \\ (-A + C)t + (-B + C + D) &= 0 , \end{aligned}$$

and you want to find A, B, C, and D that will make these equations true for <u>all</u> values of t. You can do this by setting each of the coefficients of t equal to zero and solving that new system for A, B, C, and D. Try it. What do you end up with?

6. In Step 5, you should have found $C = A$ and $B - D = A$, a system of two equations in four unknowns. This means you can choose two of the constants "at will" and solve for the other two. (Be careful - you'll spoil all your efforts if you choose $A = C = 0$.) An easy choice is $A = 1$ and $D = 0$. These choices then give $C = 1$ and $B = 1$, so that $g(t) = At + B = t + 1$ and $h(t) = Ct + D = t$. The resulting solution of system (1) is

$$\mathbf{v}(t) = ((t + 1)e^{2t}, te^{2t}) ,$$

and since it's not a constant multiple of the other solution, $\mathbf{u}(t) = (e^{2t}, e^{2t})$, the two solutions are a linearly independent set. Plug in $\mathbf{v}(t)$ to make sure it's a solution.

The general solution of system (1) is, finally,

$$C_1\mathbf{u}(t) + C_2\mathbf{v}(t) = (C_1e^{2t} + C_2(t + 1)e^{2t}, C_1e^{2t} + C_2te^{2t}) .$$

Task 1.

1. We said you'd spoil your efforts if you let $A = C = 0$ in Step 6 above. Show what happens if you do this.

2. In Step 6 above, choose the constants differently to see what happens. Let $A = 1/2$ and $B = 3/2$ and solve the system for C and D. Show that the resulting solution is just $\tfrac{1}{2}\mathbf{v} + \mathbf{u}$, a linear combination of the linearly independent pair of solutions we found in the Example.

Convince yourself that, no matter how you choose A and B, you will not get anything other than a linear combination of **u** and **v**.

Task 2. The equilibrium point $(0,0)$ for system (1) is an *improper node*. Is it stable or unstable?

Task 3.
1. Use your CAS to get a nice phase portrait for system (1). The trajectories all move away from the origin as $t \to +\infty$, so the origin is unstable. Your picture should look like this:

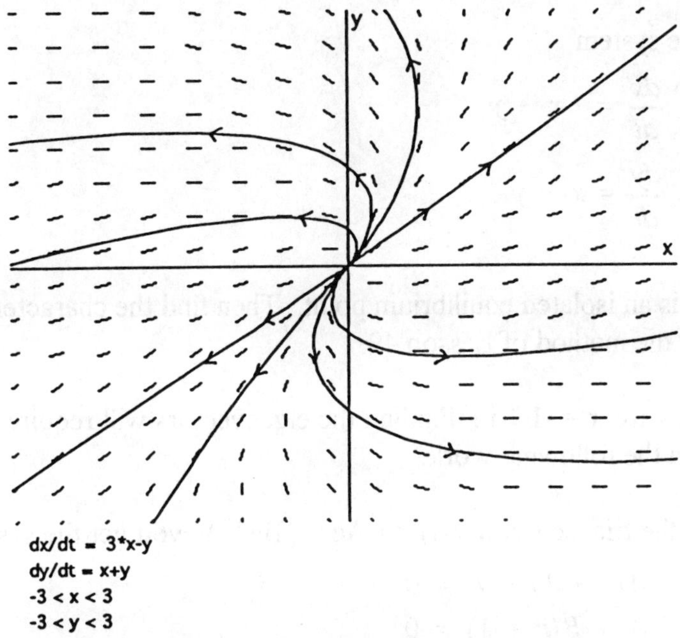

2. The trajectories should all look tangent at $(0,0)$ to the line of eigenvectors $y = x$. Do you see this behavior in your picture? Use the general solution of system (1) to form an expression for dy/dx and show that dy/dx approaches 1 as $t \to -\infty$, thus verifying the tangency claim..

Task 4. Find the solution of system (1) that goes through the point $(-3,4)$ at time $t = 0$.

Task 5. Solve the following IVP

$$\frac{dx}{dt} = x + 9y$$
$$\frac{dy}{dt} = -x - 5y,$$

$$x(0) = 1 \quad \text{and} \quad y(0) = -1,$$

a) by hand, using the method of this Lesson;
b) by using your CAS.

Task 6. Use your CAS to get a phase portrait for the above system. What is the line of tangency?

● **Complex eigenvalues and zero eigenvalues**

The next case we need to take care of is the case when the eigenvalues are complex conjugates, that is, $r = a \pm bi$. The solution method is the same as in Lesson 49, but now we'll be doing some complex number arithmetic (you might want to review Lesson 24).

Example. Look at the system

(1)
$$\frac{dx}{dt} = -x - y$$
$$\frac{dy}{dt} = x - y.$$

1. Verify that $(0,0)$ is an isolated equilibrium point. Then find the characteristic equation and eigenvalues, using the method of Lesson 49.

The eigenvalues are $r = -1 \pm i$. Finding the eigenvectors will require complex arithmetic. Carry out the details in the following work.

When you plug in the trial solution $z(t) = (\, Ae^{rt}, \, Be^{rt}\,)$, you get the system

(2)
$$A(r + 1) + B = 0$$
$$-A + B(r + 1) = 0.$$

Setting $r = -1 + i$ in the system (2) gives

(3)
$$Ai + B = 0$$
$$-A + Bi = 0.$$

These equations are equivalent (just multiply the first one by i to get the second one.) The eigenvectors are the <u>complex</u> vectors $(A, -Ai)$. That's okay; keep going. Let A be 1 and you'll have a complex solution of system (1), namely,

$$z(t) = (\, e^{(-1+i)t}, \, -i \cdot e^{(-1+i)t}\,).$$

Recall Euler's formula from Lesson 24: $e^{a+bi} = e^a [\cos(b) + i \cdot \sin(b)]$. Applying this to the solution $z(t)$ gives

$$z(t) = (\, e^{-t} [\cos(t) + i \cdot \sin(t)] \,,\, -\, i \cdot e^{-t} [\cos(t) + i \cdot \sin(t)] \,)$$

$$= (\, e^{-t} [\cos(t) + i \cdot \sin(t)] \,,\, e^{-t} [\sin(t) - i \cdot \cos(t)] \,).$$

Separate this vector into its real and imaginary parts, and write the solution as the sum of two vectors:

$$z(t) = (\ e^{-t}\cos(t)\ ,\ e^{-t}\sin(t)\ )\ +\ i \cdot (\ e^{-t}\sin(t)\ ,\ -\ e^{-t}\cos(t)\ )\ .$$

2. Show that the real part of $z(t)$,

$$u(t) = \mathfrak{Re}(z(t)) = (\ e^{-t}\cos(t)\ ,\ e^{-t}\sin(t)\ )$$

and the imaginary part,

$$v(t) = \mathfrak{Im}(z(t)) = (\ e^{-t}\sin(t)\ ,\ -\ e^{-t}\cos(t)\ )$$

are both solutions of system (1). Is u a constant multiple of v ?

The vectors $u(t)$ and $v(t)$ of Step 2 are both <u>real</u>-valued solutions, and they're linearly independent. So the general solution of system (1) is

$$(3)\qquad C_1 u(t) + C_2 v(t) = (C_1 e^{-t}\cos(t)\ +\ C_2 e^{-t}\sin(t)\ ,\ C_1 e^{-t}\sin(t)\ -\ C_2 e^{-t}\cos(t)\ )\ .$$

3. Substitute the other eigenvalue $r = -1 - i$ into equations (2) and find its eigenvectors. Then proceed as we did above to find two real-valued linearly independent solutions. Call them $s(t)$ and $w(t)$. How are s and w related to the solutions u and v we found in Step 2? So you see, the second complex eigenvalue doesn't give us anything new.

<u>Task 1.</u> Use your CAS to draw a phase portrait of system (1). The origin is a ***spiral sink*** or ***asymptotically stable focus***. All the trajectories spiral into the origin. You saw one of these pictures in Lesson 46. Here's a picture for system (1).

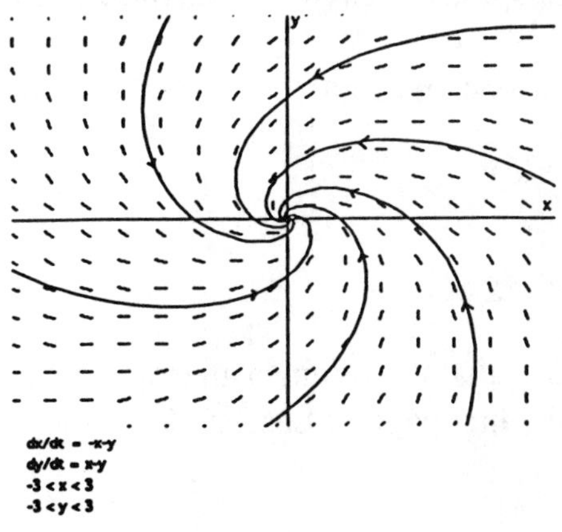

Task 2. Change the coefficients of x and y in system (1) slightly so that the eigenvalues are $r = +1 \pm i$. For this new system, the origin is a ***spiral source*** or ***unstable focus***. All the trajectories spiral away from the origin.

Task 3. The familiar system

(4)
$$\frac{dx}{dt} = y$$
$$\frac{dy}{dt} = -x$$

has, as you know from previous lessons, the two linearly independent solutions,
$\mathbf{u}(t) = (\sin(t), \cos(t))$ and $\mathbf{v}(t) = (\cos(t), -\sin(t))$. Derive these solutions using the method of this Lesson. The origin is a ***stable center*** since the trajectories are the circles $x^2 + y^2 = C^2$.

Task 4. Here's an idea that will be useful. To show you how sensitive a center is to changes in the system's coefficients, look at the system

(5)
$$\frac{dx}{dt} = ax + y$$
$$\frac{dy}{dt} = -x \; ,$$

which is just system (4) with an ax term added to its first equation.

1. Show that the eigenvalues of system (5) are $r = \dfrac{a \pm \sqrt{a^2 - 4}}{2}$.

2. Let a be a small positive number, say $a = 0.1$. Use the eigenvalues to argue that the origin is an unstable spiral source for this system.

3. Now let $a = -0.1$. What kind of equilibrium point is the origin now?

4. Finally, when $a = 0$, the origin is a stable center. So, as the coefficient a changes its value just a little bit around $a = 0$, the origin goes from being asymptotically stable to unstable, and and trajectories go from inward spirals to circles to outward spirals.

Task 5. Consider the general linear system again:

(L)
$$\frac{dx}{dt} = ax + by$$
$$\frac{dy}{dt} = cx + dy$$

Up until now we have been assuming that the origin is an <u>isolated</u> equilibrium point for system (L), that is, that $ad - bc \neq 0$ (see the first page of Lesson 49 for the discussion). Now let's see what happens if $ad - bc = 0$.

1. Show that the characteristic equation of (L) is

$$r^2 - (a + b)r + (ad - bc) = 0 .$$

2. If $ad - bc = 0$, then the characteristic equation reduces to

$$r^2 - (a + b)r = 0,$$

which has $r = 0$ as (at least) one eigenvalue. The system

$$\frac{dx}{dt} = 2x$$

(6)

$$\frac{dy}{dt} = 5x$$

is such a system. What are the equilibrium points? What are the eigenvalues? Describe the phase portrait of system (6).

Task 6. Devise coefficients a, b, c, and d for a linear system that has $r = 0$ as a repeated eigenvalue. Describe the equilibrium points and the phase portrait for your system.

You have now seen all possible behaviors for a linear homogeneous system with constant coefficients.

- **The tail of an integral**

Recall the definition of an improper integral, the kind where the interval of integration is infinite:

$$\int_{a}^{\infty} g(t)\,dt = \lim_{b \to \infty} \int_{a}^{b} g(t)\,dt$$

If the limit on the right exists, then the improper integral on the left is said to converge, and its value is that of the limit. If the limit doesn't exist, the integral diverges.

Task 1. Use the above definition to evaluate the following integrals:

1. $\displaystyle\int_{1}^{\infty} \frac{1}{t}\,dt$
2. $\displaystyle\int_{1}^{\infty} \frac{1}{t^{1.001}}\,dt$
3. $\displaystyle\int_{0}^{\infty} \frac{1}{e^{t}}\,dt$

As you can see from comparing your answers to Problems 1. and 2. of Task 1, there is a very fine line between convergence and divergence: the first integral diverges, while the second converges, yet there is hardly any difference between the funtions $1/t$ and $1/t^{1.001}$. Convergence really depends on the size of the "tail" of $g(t)$, i.e., on <u>how quickly</u> $g(t)$ approaches 0 as t goes to infinity.

Many functions $g(t)$ for which $\displaystyle\int_{a}^{\infty} g(t)\,dt$ diverges can be made to converge if they're multiplied by e^{-st} for some large enough $s > 0$. This fact will be useful when you study the Laplace Transform method of solving a DE in the next lesson.

Task 2. For each problem below, show that the improper integral from $t = 0$ to $t = \infty$ of the first one of each pair of functions diverges, while the improper integral of the second one converges, and find the value of the second one. Your CAS will be helpful here to do some of the integrations.

1. a. 1 b. e^{-st}, for any positive s

2. a. t^2 b. $t^2 e^{-st}$, for any positive s

3. a. $\sin(3t)$ b. $e^{-st}\sin(3t)$, for any positive s

You might have noticed in Task 2 that the convergent integrals you obtained were functions of the variable s ; for example, 2.b. gives

$$\int_0^\infty t^2 e^{-st}\,dt = \frac{2!}{s^3}\,, \quad s > 0 \ .$$

This will always be the case since the variable of integration, t , is evaluated between the limits t = 0 and t = ∞ , and all that's left in the end is s . You have actually built a new function of s from a function of t .

- **A first look at Laplace transforms**

You saw, when studying first order linear DEs of the form $y' + p(t)y = q(t)$, that an integrating factor for them was $\mu = e^{\int p(t)\,dt}$, and multiplication of the DE by μ enabled you to solve the equation. In a similar way, multiplying some DEs by e^{-st} and integrating over $[0,\infty)$ will also enable you to solve the DE in a rather easy way.

To make this idea clearer, we'll define the ***Laplace transform*** of a function $g(t)$ to be the following integral:

$$G(s) = \mathscr{L}(g(t)) = \int_{0}^{\infty} e^{-st} g(t)\,dt , \quad for \ \ s > 0 .$$

This improper integral will converge if $g(t)$ itself grows no faster than an exponential function; this means that $|g(t)| \leq Ke^{mt}$, where K and m are positive constants that depend upon $g(t)$. The functions we're interested in will have converging Laplace transforms - you've already seen some in Task 2 of the last lesson, where you actually computed the transforms of three functions.

Notice the notation that's used in the above definition:

$g(t)$ is the function to be transformed;

$\mathscr{L}$ is traditional notation for the Laplace transform <u>operator</u>, which turns out to be a linear operator;

$G(s)$ is the transform of $g(t)$, i.e., $\mathscr{L}$ takes in $g(t)$ and spits out $G(s)$, so that $G(s) = \mathscr{L}(g(t))$. The input function, g, is a function of t, while the output function, G, is a function of s.

So take care in distinguishing between "lower case g" and "upper case G" and their respective independent variables, t and s.

Task 1. In Task 2 of the previous lesson, you found the Laplace transforms of the three functions $g(t) = 1$, $g(t) = t^2$, and $g(t) = \sin(3t)$. Your CAS has a Laplace transform package. Use it to find the transforms of these three functions.

In the work that follows, you'll also need to know how to find the ***inverse Laplace transform*** $\mathscr{L}^{-1}$ of a function G in order to 'recover' g: If $G(s) = \mathscr{L}(g(t))$, then $g(t) = \mathscr{L}^{-1}(G(s))$. Finding $\mathscr{L}^{-1}(G(s))$ can be a formidable problem. Before the age of the CAS, people used extensive tables of Laplace transforms and various algebraic techniques to do this. Your text probably has a table of transforms.

Task 2. Use your CAS to find the inverse Laplace transforms for the following functions.

1. $\dfrac{1}{(s + 2)}$, $\quad for \ \ s > -2$

2. $\dfrac{1}{(s + 1)(s + 2)}$, $\quad for \ \ s > -1$

3. $\dfrac{s}{s^2 + 1}$, $\quad for \ \ s > 0$

4. $\dfrac{3!}{s^4}$, $\quad for \ \ s > 0$

Now we'll see how the Laplace transform can be used to solve initial value problems. Step through the following example.

Example. Let's solve the IVP

$$(1) \qquad\qquad y' + 2y = e^{-t}, \ \ y(0) = 5 \ .$$

1. This is a first order linear equation. Find its solution using the integrating factor method of Lesson 11.

You should have found the solution $y(t) = e^{-t} + 4e^{-2t}$. To get this answer using the Laplace transform method, apply $\mathscr{L}$ to both sides of equation (1) to get

$$\mathscr{L}(\, y' + 2y \,) = \ \mathscr{L}(e^{-t}) \, ,$$

and, since $\mathscr{L}$ is a linear operator, this becomes

$$(2) \qquad\qquad \mathscr{L}(y') + 2\mathscr{L}(y) = \mathscr{L}(e^{-t}) \ .$$

Fortunately, $\mathscr{L}(y')$ is related to $\mathscr{L}(y)$ in an easy way, as follows:

Property 1. If $\mathscr{L}(y(t)) = Y(s)$, then $\mathscr{L}(y'(t)) = sY(s) - y(0)$.

You can prove this property later - let's use it in equation (2) to get

$$(3) \qquad\qquad sY(s) - 5 + 2Y(s) = \mathscr{L}(e^{-t}) \, ,$$

where we've gone from the $\mathcal{L}$ notation to the Y notation and have also used the initial condition $y(0) = 5$ from the problem statement.

2. Find $\mathcal{L}(e^{-t})$, either by hand or using your CAS, and substitute your result for the right side of equation (3).

You should now have

(4) $\qquad\qquad sY(s) - 5 + 2Y(s) = 1/(s + 1)$, for $s > -1$.

Equation (4) is just an <u>algebraic</u> equation in the unknown $Y(s)$. This is one of the plusses of the Laplace method - it reduces the problem to an algebraic one.

3. Solve equation (4) for $Y(s)$.

Step 3 should result in the equation

$$(5) \qquad Y(s) = \frac{1}{(s + 1)(s + 2)} + \frac{5}{(s + 2)},$$

where $Y(s)$ is the Laplace transform of the function $y(t)$ we're after. We'll get the desired solution of our problem by applying the inverse Laplace operator $\mathcal{L}^{-1}$ to both sides of equation (5), as follows:

4. You've already found the inverse Laplace transforms for the two functions on the right side of equation (5). Look up your answers from Task 2.

5. Because $\mathcal{L}^{-1}$ is also a linear operator, you can apply it to both sides of equation (5) and break up the right side into two pieces. Use this along with your answers from Step 4 and conclude, finally, that the desired solution to the IVP (1) is

$$y(t) = \mathcal{L}^{-1}(Y(s)) = e^{-t} + 4e^{-2t}.$$

You might be wondering why we went to all this trouble to solve this IVP when we could do it much more easily the old way. The answer is that this was just an illustrative example. The real power of the Laplace transform method becomes evident when solving second order or higher IVPs, especially when there's a discontinuous forcing function, like a step function or an impulse function. These situations occur frequently in electrical and mechanical systems.

Task 3. Prove Property 1: If $\mathcal{L}(y(t)) = Y(s)$, then $\mathcal{L}(y'(t)) = sY(s) - y(0)$, by using the definition of the Laplace transform and integration by parts.

Task 4. Use the transform method to solve the following system IVP. When you need them, use your CAS to find the inverse transforms.

$$\frac{dx}{dt} = 2x - 3y \qquad x(0) = 1 \quad and \quad y(0) = 2$$

$$\frac{dy}{dt} = -2x + y \ .$$

To help you get started, apply the transform to both equations, using Property 1 and the initial conditions, and end up with the algebraic system

$$(s - 2)X(s) + 3Y(s) = 1$$
$$2X(s) + (s - 1)Y(s) = 2 \ .$$

Solve this system algebraically for $X(s)$ and $Y(s)$ and then find their inverse transforms. Your final answer should be

$$x(t) = \frac{8}{5}e^{-t} - \frac{3}{5}e^{4t} \quad and \quad y(t) = \frac{8}{5}e^{-t} + \frac{2}{5}e^{4t} \ .$$

Check your answer by plugging it into the original system.

•   **The Laplace transform: A second order example**

As another example, let's use the Laplace transform method on a forced mass-spring problem. We'll need to know what $\mathcal{L}(y'')$ is; see Property 2 below. For completeness, we'll restate Property 1 of the last lesson, too.

Property 1. If $\mathcal{L}(y(t)) = Y(s)$, then $\mathcal{L}(y'(t)) = sY(s) - y(0)$.

Property 2. If $\mathcal{L}(y(t)) = Y(s)$, then $\mathcal{L}(y''(t)) = s^2Y(s) - s\,y(0) - y'(0)$.

Work through the following example.

Example. Consider the IVP for a forced mass-spring system given by

(1) $$y'' + 3y' + 2y = \sin(2t), \quad \text{where } y(0) = 1 \text{ and } y'(0) = -3.$$

If you were to solve this problem the "old" way, you would first have to find the general solution of the associated homogeneous DE, then find a particular solution of equation (1) using either the method of undetermined coefficients or variation of parameters, then find the general solution of (1), and, finally, solve for the constants using the initial conditions.

The Laplace transform method neatly does everything at once. The hard part will be finding the inverse transforms - but we can make the CAS do that.

1. Use your CAS to solve the IVP (1), just as you did in Task 4 of Lesson 32, so you'll know what solution we're aiming for.

2. Now apply the Laplace operator $\mathcal{L}$ to the DE, making use of Properties 1 and 2 above, and the initial conditions. What do you get for $Y(s)$?

You should have derived the equation

$$s^2Y(s) - s + 3 + 3sY(s) - 3 + 2Y(s) = 2/(s^2 + 4),$$

which simplifies to

(2) $$Y(s) = \frac{s^3 + 4s + 2}{(s^2 + 4)(s^2 + 3s + 2)}.$$

If you were to use the table look-up method to find the inverse transform of $Y(s)$, you would first have to use the method of partial fractions to write the right side of equation (2) as the sum of three fractions, then look up the inverse transform for each fraction. But we'll just use the computer.

3. Use your CAS to find the inverse transform of $Y(s)$ given above. The result is the desired solution to the IVP and should match the solution you found in Step 1, namely,

$$y(t) = -\frac{3}{20}\cos(2t) - \frac{1}{20}\sin(2t) - \frac{3}{5}e^{-t} + \frac{7}{4}e^{-2t} .$$

__Task 1.__ Solve the IVP

$$y'' - 3y' + 2y = \cos(2t), \quad \text{where } y(0) = -3 \text{ and } y'(0) = 5$$

in two ways:

a) using your CAS directly;
b) using the Laplace transform method.

There is much, much more to be said about this method. It becomes particularly powerful when the forcing function $f(t)$ is piecewise continuous. See your text for details.

APPENDIX A

MAPLE AND MATHEMATICA SYNTAX

Veryifying solutions and using your CAS to find analytic solutions

MAPLE:

Task 1. We'll illustrate how to verify that the family $y = \exp(t)/(\exp(t) + C)$ is the general solution of equation 7, namely, dy/dt = y - y^2 .

```
> y:=exp(t)/(exp(t)+C);   #Define the solution family.
```

$$y := \frac{e^t}{e^t + C}$$

```
> simplify(y-y^2);   #Simplify the rhs of the differential equation.
```

$$\frac{e^t C}{\left(e^t + C\right)^2}$$

```
> simplify(diff(y,t));   #Find the derivative and simplify it.
```

$$\frac{e^t C}{\left(e^t + C\right)^2}$$

Comparing the last two outputs above shows that you have a solution.

Task 2. To solve a DE for analytic solutions, we'll again illustrate using equation 7.

```
> y:='y': #Clear y to use it again.
> eqn:=diff(y(t),t)=y(t) - y(t)^2;   #Define the DE.
```

$$eqn := \frac{\partial}{\partial t} y(t) = y(t) - y(t)^2$$

```
> soln:=dsolve(eqn,y(t)); #Use the 'dsolve' command.
```

$$soln := \frac{1}{y(t)} = 1 + e^{(-t)} \_C1$$

Notice that the solution is given implicitly. We can ask for an explicit form and, if possible, Maple will provide it. Note also Maple's form of the integration constant: _C1.

```
> soln2:=dsolve(eqn,y(t),explicit);
```

$$soln2 := y(t) = \frac{1}{1 + e^{(-t)} \_C1}$$

Now you can do some algebra by hand to verify that this is the same solution as given in Lesson 3. Notice that Maple did not give the trivial solution, $y(t) = 0$. It's a special case that you have to find by hand.

```
> ?dsolve # Look at syntax and options for the dsolve command
```

MATHEMATICA:

Task 1. We'll illustrate how to verify that the family $y = \exp(t)/(\exp(t) + C)$ is the general solution of equation 7, namely, $dy/dt = y - y^2$.

```
Clear[y,t,c]  (*It's a good idea to reset variables you will use.*)

y=Exp[t]/(Exp[t]+c)  (*Define the solution family*)

Simplify[y-y^2]  (*Simplify rhs of de*)

Simplify[D[y,t]]  (*Find derivative amd simplify*)
```


Task 2. To solve a DE for analytic solutions, we'll again illustrate using equation 7.

```
Clear[y]  (*Unassign y*)

eqn=y'[t]==y[t]-y[t]^2 (*Define the de. Note the double ==*)

DSolve[eqn,y[t],t] (*Solve the de. There are two solutions. Note the arbitrary constant*)

soln=DSolve[eqn,y[t],t] (*Or solve this way*)

y1[t_]=y[t]/.First[soln] (* Assign y1 to be the first solution. Be sure to use =, not :=, here*)

y1[t] (*Look at y1*)

y1[0] (* Evaluate y1 at t=0 *)

?DSolve (*Look at syntax for the DSolve command*)
```

Using your CAS to plot families of solutions
Solving initial value problems

MAPLE:

<u>Task 1.</u> Since we already know the solutions, we can easily plot them by forming a set of solutions with the C-values we want, and then plot the set. We'll illustrate the method with the solutions of Equation 7.

```
> with(plots):
> y:=exp(t)/(exp(t)+C);   #Define the family of solutions.
```

$$y := \frac{e^t}{e^t + C}$$

```
> toplot:={seq(subs(C=i,y),i=-2..2)};   #Create a set of the sequence
  of solutions.
```

$$toplot := \left\{ 1, \frac{e^t}{e^t - 2}, \frac{e^t}{e^t + 2}, \frac{e^t}{e^t - 1}, \frac{e^t}{e^t + 1} \right\}$$

```
> plot(toplot,t=-5..5,-5..5);
```

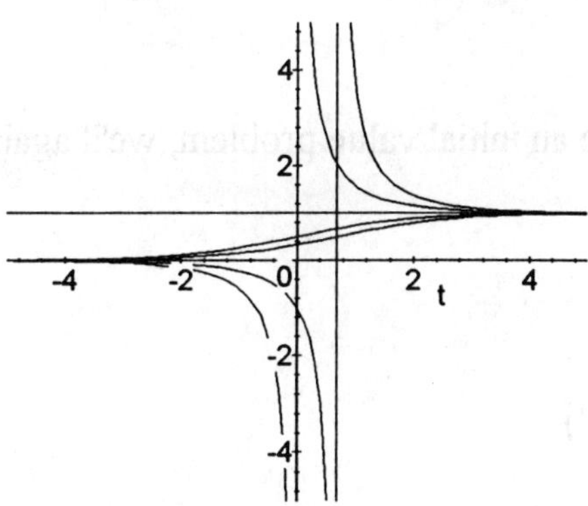

<u>Task 3.</u> To illustrate how to solve an initial value problem, we'll again use equation 7 and its initial condition $y(0) = 0.5$.

```
> y:='y':   # Reset y.
> eqn:=diff(y(t),t)=y(t) - y(t)^2;   # Define the DE.
```

$$eqn := \frac{\partial}{\partial t} y(t) = y(t) - y(t)^2$$

```
> soln:=dsolve({eqn,y(0)=0.5},y(t),explicit=true);  # Include the
  initial condition in the dsolve command.
```

$$soln := y(t) = \frac{1}{1 + 1.\, e^{(-t)}}$$

We can check the answer as follows:

```
> y:=rhs(soln);   #Set  y  equal  to  the  right-hand  side  of  the  soln.
```

$$y := \frac{1}{1 + 1.\, e^{(-t)}}$$

```
> evalf(subs(t=0,y));   #Evaluate  y  at  t=0.   The  answer  is  correct.
                          .5000000000
```

MATHEMATICA:

Task 1. Since we already know the solutions, we can easily plot them by forming a set of solutions with the C-values we want, and then plot the set. We'll illustrate the method with the solutions of Equation 7.

```
y[t_,c_]=Exp[t]/(Exp[t]+c)  (*Define the function with parameter c*)

toplot:=Table[y[t,c],{c,-2,2}]  (*Define a list of functions*)

Plot[Evaluate[toplot],{t,-5,5}]  (*Plot the functions*)
```

Task 3. To illustrate how to solve an initial value problem, we'll again use equation 7 and its initial condition $y(0) = 0.5$.

```
Clear[y,t, eqn]  (* Reset variables *)

eqn=y'[t]==y[t]-y[t]^2  (* Define de *)

soln=DSolve[{eqn,y[0]==.5},y[t],t]  (* Solve with initial condition *)

y1[t_]=y[t]/.First[soln]  (* Define y1 to be the solution *)

y1[t]  (* Look at y1 *)

y1[0]  (* Evaluate y1 at t=0 *)
```

Slope fields

MAPLE:

<u>Tasks 1. & 2.</u> You can use the following commands to draw slope fields. We'll illustrate them for equation 7.

```
> with(DEtools):   # Call up the DEtools package.

> y:='y':  # Reset y, if you need to.
> eqn:=diff(y(t),t)=y(t) - y(t)^2;   # Define the DE.
```

$$eqn := \frac{\partial}{\partial t} y(t) = y(t) - y(t)^2$$

```
> dfieldplot(eqn,y(t),t=-3..3,y=-3..3);   # Plot the direction field.
```

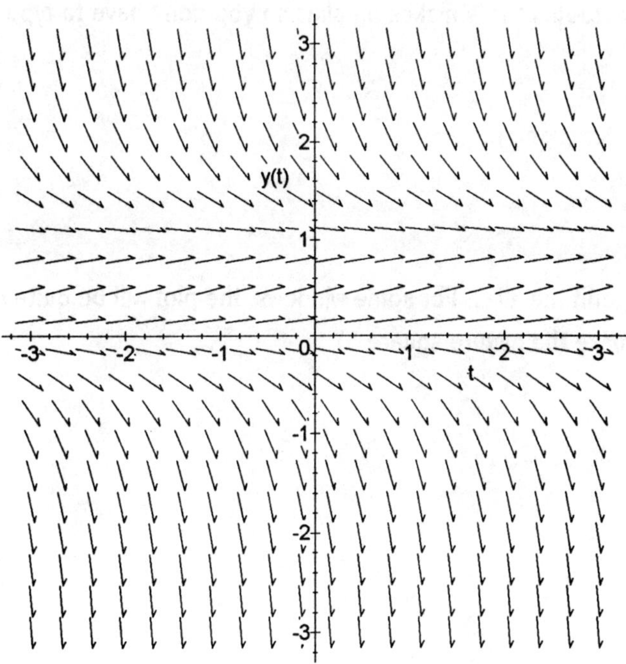

MATHEMATICA:

Tasks 1. & 2. You can use the following commands to draw slope fields. We'll illustrate them for equation 7.

<<Graphics`PlotField` (* Load graphics package *)

m[t_,y_]=y-y^2 (* Define the function; solve for dy/dt. *)

PlotVectorField[{1,m[t,y]},{t,-3,3},{y,-3,3}] (* Draw the vector field. Note the arrows are different lengths. *)

Options[PlotVectorField] (* Look at the options for the command. The ScaleFunction command normalizes the arrow lengths. *)

dfield:=PlotVectorField[{1,m[t,y]},{t,-3,3},{y,-3,3},

ScaleFunction->(1&),Axes->True] (* This makes an alias so you don't have to type so much each time. Note the := *)

m[t_,y_]=y (* Define a new function *)

dfield (* Use the alias *)

(* To change the plot range, edit the alias. For some windows, the plot will be quite narrow - use the option AspectRatio->1 to make the picture square *)

The inaccuracy of Euler's method

MAPLE:

We'll illustrate the steps of Lesson 7 using the initial value problem
$$dy/dt = y , y(0) = 1 , \text{ on the t-interval } [0,1] .$$
You can adapt the following commands to do the Example given in Lesson 7.

```
> eqn:=diff(y(t),t)=y(t); #Define the DE
```

$$eqn := \frac{\partial}{\partial t} y(t) = y(t)$$

Maple has several different numerical methods with which to solve a DE. The Euler method is called the classical method. Here's the command; notice how the stepsize is specified:

```
> sol:=dsolve({eqn,y(0)=1},y(t),type=numeric,method=classical,stepsi
  ze=0.1);
```

$$sol := \mathbf{proc}(x\_classical) \dots \mathbf{end}$$

The above output line just indicates that the procedure was done successfully. To see what's in the variable 'sol', you can just ask for a value; let's ask for the value when t = 0.5:

```
> sol(0.5);
```

$$[t = .5, y(t) = 1.610510000000000]$$

Notice that Maple's response was a list of two values. We can't use that format for the plot we want, so we'll turn it into something useful. Notice the use of "Y" and "c" in the following command; we don't want to spoin "y" and "t", since they're already being used as variables and we need to preserve them.

```
> Y:=c->subs(sol(c),y(t));
```

$$Y := c \rightarrow subs(sol(c), y(t))$$

Now our solution values are in the function Y. We can print the first six Y values using a simple for loop, as follows:

```
> for i from 0 to 5 do print(Y(i/10.)) od;
```

$$1.$$
$$1.100000000000000$$
$$1.210000000000000$$
$$1.331000000000000$$
$$1.464100000000000$$

$$1.610510000000000$$

We can now plot the first eleven points of the Euler solution, as follows:

```
> with(plots):   # Call up the plots package.
> pointplot({seq([i/10.,Y(i/10)],i=0..10)},style=point,symbol=circle
  ,color=red);
```

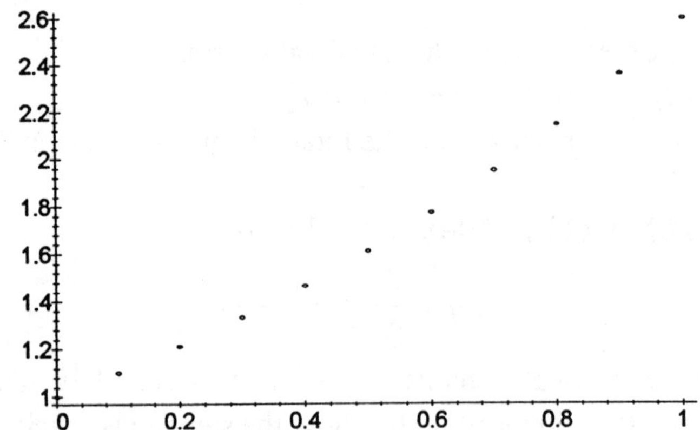

Now, let's ask Maple to solve the IVP exactly:

```
> sol2:=dsolve({eqn,y(0)=1},y(t));
```

$$sol2 := y(t) = \mathbf{e}^t$$

Just as we did with the numerical solution, we'll get this solution into a more useful form and then look at six values for the same t-values as before:

```
> Y2:=rhs(sol2);
```

$$Y2 := \mathbf{e}^t$$

```
> for i from 0 to 5 do print(evalf(subs(t=i/10.,Y2))) od;
```

$$1.$$
$$1.105170918$$
$$1.221402758$$
$$1.349858808$$
$$1.491824698$$
$$1.648721271$$

Compare this list with the previous one to see the difference between the actual solution and the Euler numerical solution.

Now, let's plot both solutions on the same axes. We'll define the two plots and store them in the variables p1 and p2 , and then display both plots together. Notice the use of semi-colons to suppress the output while we're defining the plots.

```
> p1:=pointplot({seq([i/10.,Y(i/10)],i=0..10)},style=point,symbol=ci
```

```
      rcle,color=red):
> p2:=plot(Y2,t=0..1,color=black):
> display({p1,p2});
```

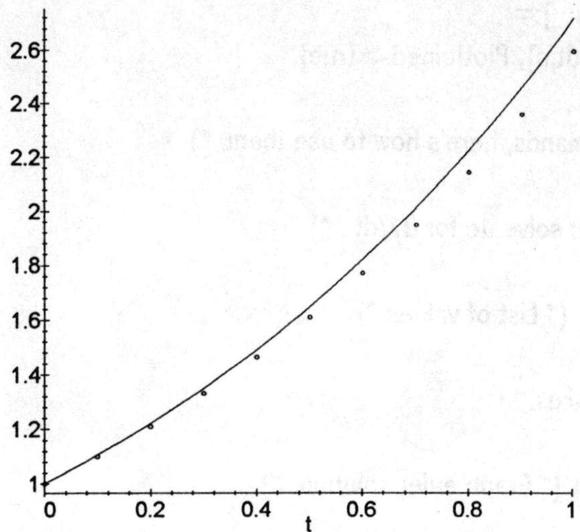

MATHEMATICA:

We'll illustrate the steps of Lesson 7 using the initial value problem

$$dy/dt = y \, , \ y(0) = 1 \, , \text{ on the t-interval } [0,1] \, .$$

You can adapt the following commands to do the Example given in Lesson 7.

Mathematica doesn't have an Euler method routine, so you'll have to define your own, which is done for you below. You just have to type in the commands. First, execute the commands exactly as they appear below to see how they work. The solution is the familiar exponential function, exp(t). Then try them on the problem in Lesson 7.

```
eqn:=y'[t]==y[t]  (* Define the de *)
```

(* Below is a sequence of four commands which allow you to execute Euler's method, and either list the results or graph them. If you can save these commands, do so, then load them when you need them. *)

```
estep[func_,{t_,y_},dt_]:=

   Block[ {nf,slope},
          nf[at_,ay_]:=N[func[at,ay]];
          slope = nf[t,y];
          {t+dt , y + slope*dt}
        ]
euler[func_,{t0_,y0_},dt_,n_]:=
   NestList[estep[func,#,dt]&,{t0,y0},n]
```

```
eulernum[func_,{t0_,y0_},dt_,n_]:=
    TableForm[euler[func,{t0,y0},dt,n]]

eulerplot[func_,{t0_,y0_},dt_,n_]:=
    ListPlot[euler[func,{t0,y0},dt,n], PlotJoined->True]

(* Having executed these commands, here's how to use them. *)

m[t_,y_]:=y  (* Define function: solve de for dy/dt . *)

elist=eulernum[m,{0,1},.1,10]  (* List of values *)

N[elist,10]  (* More decimal places. *)

eplot=eulerplot[m,{0,1},.1,10]  (* Graph euler solution. *)

soln:=DSolve[{eqn,y[0]==1},y[t],t]  (* Solve the de exactly *)

y1[t_]=y[t]/.First[soln]  (* Assign y1 to be the solution *)

sollist=Table[{t,y1[t]},{t,0,1,.1}]  (* Table of exact values. *)

TableForm[sollist]  (* More easily read. *)

ListPlot[sollist,PlotJoined->True,PlotStyle->RGBColor[1,0,0]]
  (* Plot exact solution values as a list. *)

solplot=Plot[y1[t],{t,0,1}, PlotStyle->RGBColor[0,0,1]]
  (* Or plot the solution in the usual way. *)

Show[eplot,solplot]  (* Compare the graphs. *)
```

A harder mixing problem

MAPLE:

Tasks 1 - 4. We'll illustrate computer answers to Tasks 1 - 4 using a slightly different IVP, namely,

$$dx/dt = 1.0 - 7\,x/(100 - 2t),\quad x(0) = 1.$$

You should be able to produce results similar to those below for the problem in the Lesson.

```
> eqn:=diff(x(t),t) = 1.0 - 7*x(t)/(100 - 2*t);   # Enter the DE.
```

$$eqn := \frac{\partial}{\partial t}\,x(t) = 1.0 - 7\,\frac{x(t)}{100 - 2t}$$

```
> Digits:=3:
> sol:=dsolve({eqn,x(0)=1},x(t),explicit);   # Solve the IVP.
```

$$sol := x(t) = 20. - 1.90\sqrt{100. - 2.\,t} + .114\sqrt{100. - 2.\,t}\,t - .00228\sqrt{100. - 2.\,t}\,t^2 - .400\,t$$
$$+ .0000152\sqrt{100. - 2.\,t}\,t^3$$

In order to work with this solution, we need to store it in a variable, X. (Don't use lower case x, as that's reserved for the variable in the DE and you'll spoil the DE if you want to use it again.)

```
> X:=rhs(sol);
```

$$X := 20. - 1.90\sqrt{100. - 2.\,t} + .114\sqrt{100. - 2.\,t}\,t - .00228\sqrt{100. - 2.\,t}\,t^2 - .400\,t$$
$$+ .0000152\sqrt{100. - 2.\,t}\,t^3$$

```
> plot(X,t=0..70,y=0..10);
```

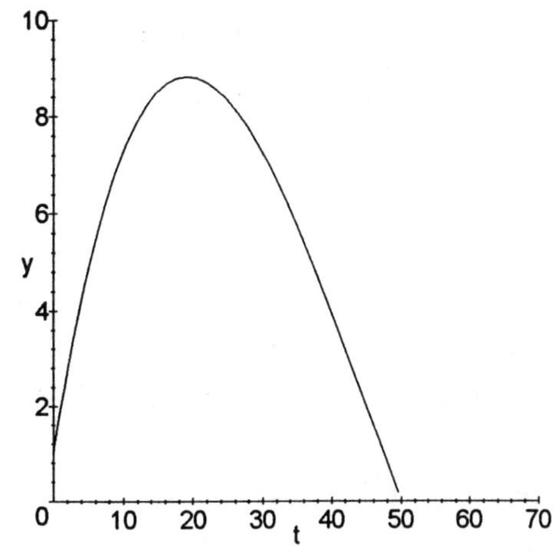

```
>
> conc:=X/(100-2*t); # Define the concentration for Task 3.
```

$$conc := (20. - 1.90\sqrt{100. - 2.\,t} + .114\sqrt{100. - 2.\,t}\,t - .00228\sqrt{100. - 2.\,t}\,t^2 - .400\,t$$
$$+ .0000152\sqrt{100. - 2.\,t}\,t^3)/(100 - 2\,t)$$

```
> limit(conc,t=50);   # The answer to Task 3.
```

$$.200$$

It would now be helpful to have graphs of the concentration and the constant function 0.1 on the same axes, so you can see what's happening.

```
> plot({conc,0.1},t=0..50);
```

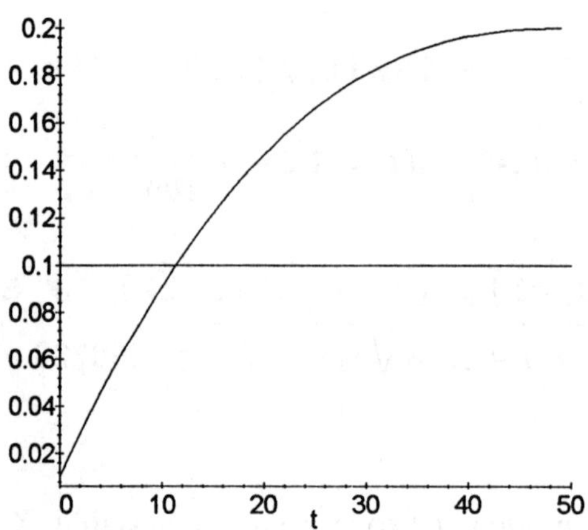

The following 'fsolve' command uses numerical approximation techniques to find a solution. Very often, you have to give it a t-range on which to search, but not this time. Type ?fsolve to see the help screen and examples for this command.

```
> fsolve(conc=0.1,t); # The answer to Task 4, in minutes.
```

11.3

MATHEMATICA:

Tasks 1 - 4. We'll illustrate computer answers to Tasks 1 - 4 using a slightly different IVP, namely,

$$dx/dt = 1.0 - 7\,x/(100 - 2t)\,, \quad x(0) = 1.$$

You should be able to emulate the results below for the problem in the Lesson.

```
eqn:=x'[t]==1.0-7*x[t]/(100-2*t) (* Define the de. The
    decimal point causes all numbers to be floating
    point, rather than exact. *)

soln=DSolve[{eqn,x[0]==1},x[t],t] (* Solve the IVP. *)
```

$$\{\{x[t] \to 20. - 0.000021496\,(50 - t)^{7/2} - 0.4\,t\}\}$$

```
x1[t_]=x[t]/.First[soln] (* Define x1 to be the solution. *)
```

$$20. - 0.000021496\,(50 - t)^{7/2} - 0.4\,t$$

**Plot[x1[t],{t,0,70}] (* Graph solution. Note error
    messages becauses the function is undefined past 50. *)**

Plot::plnr: CompiledFunction[{t}, <<1>>, -CompiledCode-][t]
    is not a machine-size real number at t = 52.5.

Plot::plnr: CompiledFunction[{t}, <<1>>, -CompiledCode-][t]
    is not a machine-size real number at t = 51.0417.

Plot::plnr: CompiledFunction[{t}, <<1>>, -CompiledCode-][t]
    is not a machine-size real number at t = 50.3125.

General::stop:
    Further output of Plot::plnr
        will be suppressed during this calculation.

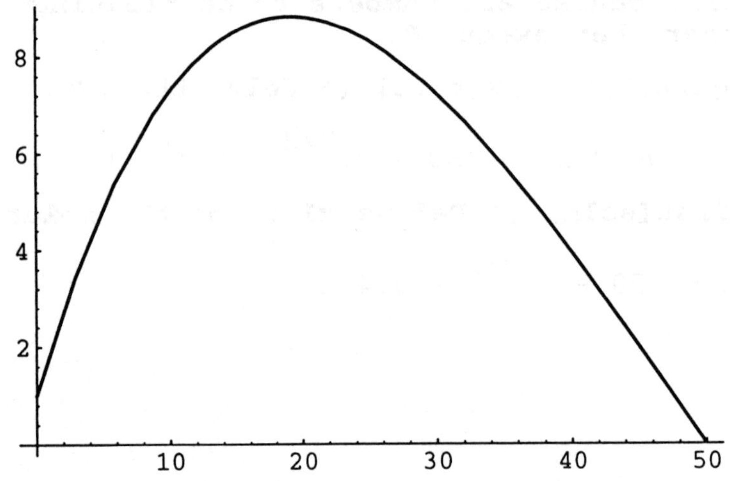

conc[t_]:=x1[t]/(100-2*t) (* Define the concentration. *)

conc[t] (* Look at the function. *)

$$\frac{20. - 0.000021496\ (50 - t)^{7/2} - 0.4\ t}{100 - 2\ t}$$

Plot[conc[t],{t,0,50}] (* Graph the concentration. Note
    the error messages at t=50. *)

                                    1
Power::infy: Infinite expression -- encountered.
                                    0.

Infinity::indet:
    Indeterminate expression 0. ComplexInfinity encountered.

                                    1
Power::infy: Infinite expression -- encountered.
                                    0.

Infinity::indet:
    Indeterminate expression 0. ComplexInfinity encountered.

Plot::plnr: CompiledFunction[{t}, <<1>>, -CompiledCode-][t]
    is not a machine-size real number at t = 50..

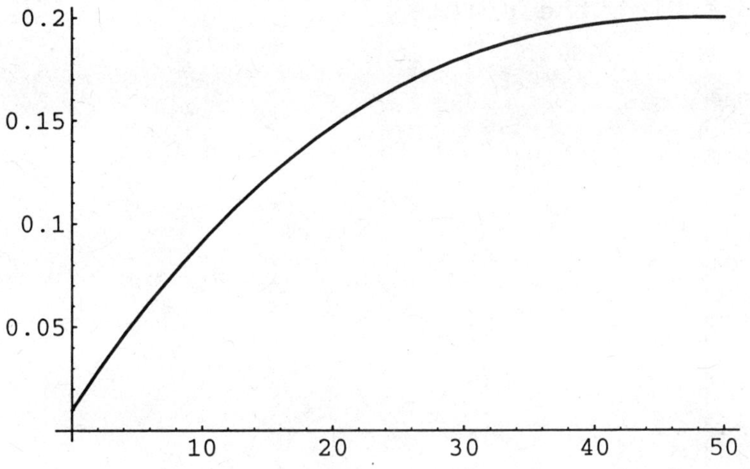

Limit[conc[t],t->50] (* The answer to Task 3. *)

0.2

Solve[conc[t]==0.1,t] (* The answer to Task 4. Ignore the
    complex solutions. *)

{{t -> 11.3216}, {t -> 81.2915 - 22.7346 I},
    {t -> 81.2915 + 22.7346 I}}

The Malthusian population growth model using laboratory data.

MAPLE:

Questions 2 and 6. Plotting the data points and the solution curve.

First, make a list of the data points; then plot them.

```
> A:=[[0,0.02],[.5,.024],[1,.031],[2,.043],[3,.062],[4,.08],[6,.163]
  ];
```

$$A := [[0, .02], [.5, .024], [1, .031], [2, .043], [3, .062], [4, .08], [6, .163]]$$

```
> plot(A,style=point); # Plot the points.
```

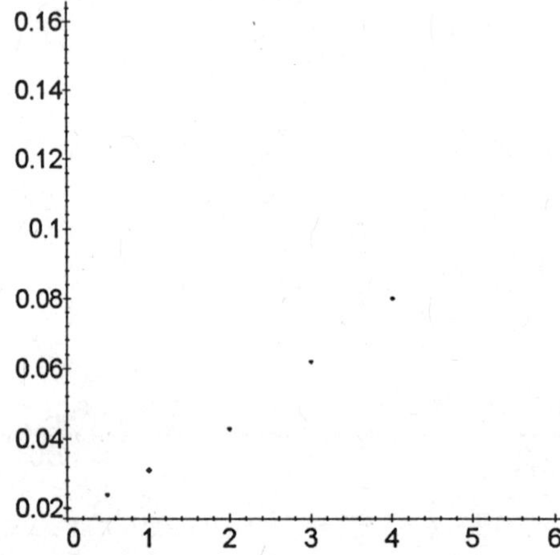

Now define your solution from Task 5, and plot it and the points.

```
> p:=0.02*exp(0.377*t); # Define the solution curve
```

$$p := .02\, e^{(.377\, t)}$$

```
> with(plots):
> p1:=plot(A,style=point): # Plot the points and save the plot.
> p2:=plot(p,t=0..10): #Plot the curve and save the plot.
> display({p1,p2}); #Show the two plots together.
```

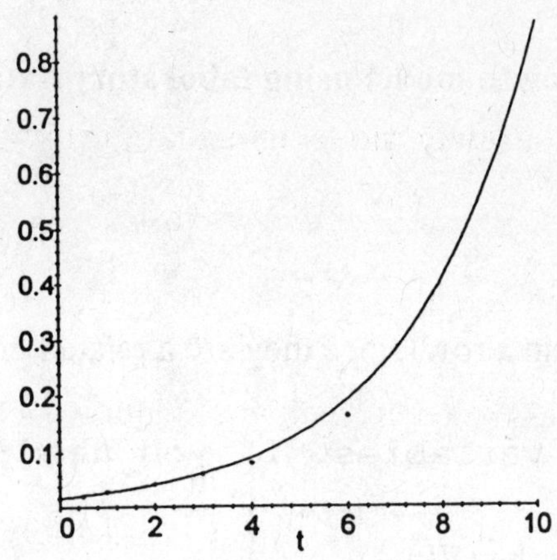

MATHEMATICA:

<u>Questions 2 and 6</u>. Plotting the data points and the solution curve.

a={{0,.02},{.5,.024},{1,.031},{2,.043},{3,.062},{4,.08},
{6,.163}} (* Define the list of points. *)

ListPlot[a] (* Plot the points. *)

lplot=ListPlot[a,PlotRange->{0,.2},

PlotStyle->PointSize[.02]] (* Make a nicer plot. *)

p[t_]:=.02*Exp[.377*t] (* Define the solution from Tasks 4 and 5. *)

solplot=Plot[p[t],{t,0,10}] (* Graph the solution. *)

Show[lplot,solplot] (* Show both graphs. *)

The logistic population growth model using laboratory data

MAPLE:

We shall do Questions 1 - 5 all in a row, since they are a related group of problems.

```
> restart;#reset all variables, if you need to.
```

Question 1. Have Maple solve the IVP:

```
> eqn:=diff(p(t),t)=a*p(t)-b*p(t)^2; #Enter the DE
```

$$eqn := \frac{\partial}{\partial t}\,\mathrm{p}(t) = a\,\mathrm{p}(t) - b\,\mathrm{p}(t)^2$$

```
> sol:=dsolve({eqn,p(0)=p0},p(t),explicit);
```

$$sol := \mathrm{p}(t) = \frac{a}{b - \dfrac{e^{(-a\,t)}\,(-a + p0\,b)}{p0}}$$

```
> sol2:=simplify(rhs(sol));# Simplify and store the solution
```

$$sol2 := -\frac{a\,p0}{-p0\,b - e^{(-a\,t)}\,a + e^{(-a\,t)}\,p0\,b}$$

Question 3. To solve for a and b , first enter their equations, then substitute appropriate values for the data points. We'll use the values of a and b to fit a solution to the data points.

```
> a:=(1/t1)*ln(p2*(p1-p0)/(p0*(p2-p1)));
```

$$a := \frac{\ln\!\left(\dfrac{p2\,(p1 - p0)}{p0\,(p2 - p1)}\right)}{t1}$$

```
> b:=(a/p1)*(p1*p1-p0*p2)/(p1*p2-2*p0*p2+p0*p1);
```

$$b := \frac{\ln\!\left(\dfrac{p2\,(p1 - p0)}{p0\,(p2 - p1)}\right)(p1^2 - p0\,p2)}{t1\,p1\,(p1\,p2 - 2\,p0\,p2 + p0\,p1)}$$

```
> Digits:=4:
> a:=evalf(subs(t1=3,p0=.02,p1=.062,p2=.163,a)); # Find  a
```

$$a := .4070$$

```
> b:=evalf(subs(p0=.02,t1=3,p1=.062,p2=.163,b)); # Find  b
```

$$b := .7943$$

```
> sol3:=subs(p0=.02,sol2); # p(t) for Question 3
```

$$sol3 := -\frac{.008140}{-.01589 - .3911\, e^{(-.4070\, t)}}$$

Question 4. Now we'll input the data points and then plot them and the solution together.

```
> A:=[[0,.02],[.5,.024],[1,.031],[2,.043],[3,.062],[4,.08],[6,.163]]
  ;
```

$$A := [[0, .02], [.5, .024], [1, .031], [2, .043], [3, .062], [4, .08], [6, .163]]$$

```
> with(plots):
> p1:=plot(A,style=point,color=red):# Plot pts. and save
> p2:=plot(sol3,t=0..12,color=blue):# Plot soln. and save
> display({p1,p2}); # Display both plots together
```

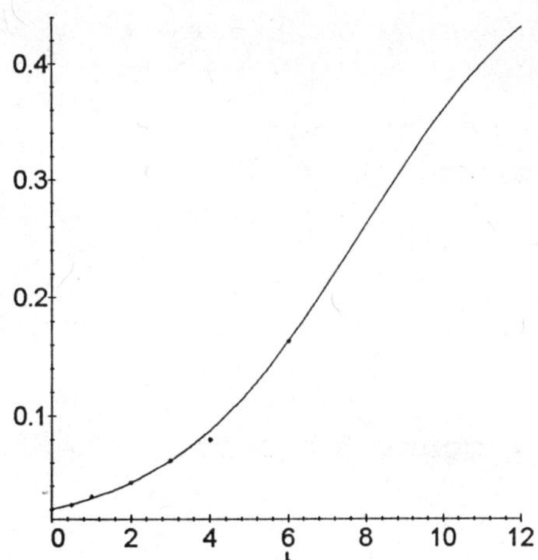

Question 5. You may answer these questions with the following commands:

```
> evalf(subs(t=12,sol3)); # Predicted population at  t = 12
```

$$.4318$$

```
> limit(sol3,t=infinity); # Behavior for large t
```

$$.5123$$

```
> evalf(a/b);
```

$$.5124$$

A 16-2

MATHEMATICA:

We shall do Questions 1-5 all in a row, since they are a related group of problems.

In[1]:=
```
eqn:=p'[t]==a*p[t]-b*p[t]^2 (* Define the de *)
```

In[2]:=
```
soln=DSolve[eqn,p[t],t] (* Solve the de *)
```

Out[2]=

$$\left\{\left\{p[t] \; \rightarrow \; \frac{a \; E^{a \; t}}{b \; E^{a \; t} \; + \; a \; C[1]}\right\}, \; \{p[t] \; \rightarrow \; 0\}\right\}$$

In[3]:=
```
p1[t_]=p[t]/.First[soln] (* Define the solution *)
```

Out[3]=

$$\frac{a \; E^{a \; t}}{b \; E^{a \; t} \; + \; a \; C[1]}$$

In[4]:=
```
const=Solve[p1[0]==p0,C[1]] (* Solve for the constant
     C[1] in terms of p0 *)
```

Out[4]=

$$\left\{\left\{C[1] \; \rightarrow \; \frac{a \; - \; b \; p0}{a \; p0}\right\}\right\}$$

In[5]:=
```
p[t_] = p1[t] /. const (* Define the solution in terms
     of p0 *)
```

Out[5]=

$$\left\{\frac{a \; E^{a \; t}}{b \; E^{a \; t} \; + \; \dfrac{a \; - \; b \; p0}{p0}}\right\}$$

```
(* Task 3 *)
```

In[6]:=
```
a = (1/t1)*(Log[p2*(p1-p0)/(p0*(p2-p1))]) (* Define the
     parameters *)
```

Out[6]=

$$\frac{Log\left[\dfrac{(-p0 \; + \; p1) \; p2}{p0 \; (-p1 \; + \; p2)}\right]}{t1}$$

In[7]:=
```
b = (a/p1)*(p1*p1-p0*p2)/(p1*p2-2*p0*p2+p0*p1)
```

Out[7]=

$$\frac{(p1^2 - p0\ p2)\ Log[\frac{(-p0 + p1)\ p2}{p0\ (-p1 + p2)}]}{p1\ (p0\ p1 - 2\ p0\ p2 + p1\ p2)\ t1}$$

In[8]:=
```
t1=3 (* Enter values for t1, p0, p1, p2 *)
```

Out[8]=
```
3
```

In[9]:=
```
p0=.02
```

Out[9]=
```
0.02
```

In[10]:=
```
p1=.062
```

Out[10]=
```
0.062
```

In[11]:=
```
p2=.163
```

Out[11]=
```
0.163
```

```
{t1,p0,p1,p2}={3,.02,.062,.163} (* Alternate syntax for
     entering lots of values *)
```

Out[22]=
```
{3, 0.02, 0.062, 0.163}
```

In[27]:=
```
{a,b} (* Observe the values of a and b *)
```

Out[27]=
```
{0.406856, 0.794098}
```

In[14]:=
```
p[t] (* Observe the solution with values defined *)
```

Out[14]=

$$\{\frac{0.406856\ E^{0.406856\ t}}{19.5487 + 0.794098\ E^{0.406856\ t}}\}$$

In[15]:=
```
solplot=Plot[p[t],{t,0,12}]  (* Plot the solution *)
```

In[16]:=
```
data={{0,.02},{.5,.024},{1,.031},{2,.043},{3,.062},{4,.08},
    {6,.163}} (* Enter the data - don't name it a again! *)
```

Out[16]=
```
{{0, 0.02}, {0.5, 0.024}, {1, 0.031}, {2, 0.043}, {3, 0.062},
  {4, 0.08}, {6, 0.163}}
```

In[17]:=
```
lplot=ListPlot[data,PlotRange->{0,.2},
    PlotStyle->PointSize[.02]] (* Plot the data *)
```

In[18]:=
    Show[solplot,lplot] (* Compare both graphs *)

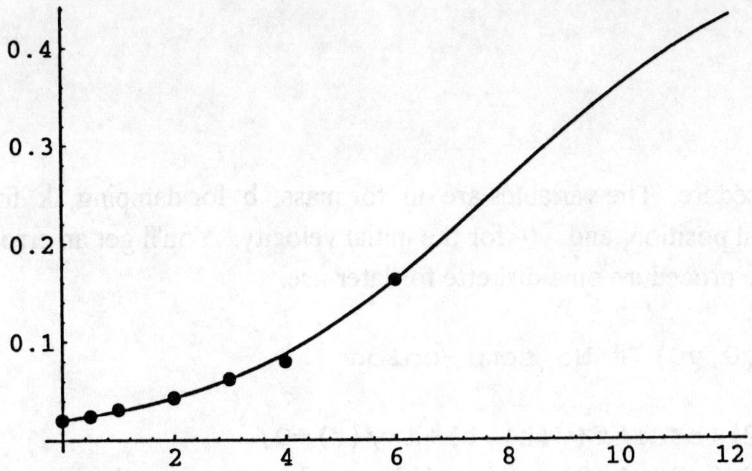

    (* Task 5 *)

In[19]:=
    p[12] (* Evaluate the predicted population at t=12 *)

Out[19]=
    {0.431781}

In[20]:=
    Limit[p[t],t->Infinity] (* Eventual population *)

Out[20]=
    {0.512349}

In[21]:=
    a/b (* Observe the value of a/b *)

Out[21]=
    0.512349

Entering the 'spring' program into your CAS

MAPLE:

Task 1. Here is the 'spring' procedure. The variables are m for mass, b for damping, k for the spring constant, y0 for the initial position, and v0 for the initial velocity. You'll get an error message until you type in 'end;'. Save the procedure on a diskette for later use.

```
> spring:=proc(m,b,k,y0,v0)  # No semi-colon
> local eqn, t, y, s;
> eqn:=m*diff(y(t),t$2)+b*diff(y(t),t)+k*y(t)=0;
> s:=dsolve({eqn,y(0)=y0,D(y)(0)=v0},y(t),explicit=true);
> s:=subs(y(t)=y,rhs(s));
> plot(s,t=0..15);  # You may change this domain if necessary
> end;
```

$spring := \mathbf{proc}(m, b, k, y0, v0)$

$\mathbf{local}\ eqn, t, y, s;$

$\quad eqn := m*\text{diff}(y(t), t \$ 2) + b*\text{diff}(y(t), t) + k*y(t) = 0;$

$\quad s := \text{dsolve}(\{y(0) = y0, D(y)(0) = v0, eqn\}, y(t), explicit = true);$

$\quad s := \text{subs}(y(t) = y, \text{rhs}(s));$

$\quad \text{plot}(s, t = 0 .. 15)$

$\mathbf{end}$

Task 2. To test the procedure, type:

```
> spring(1,0,1,1,0);
```

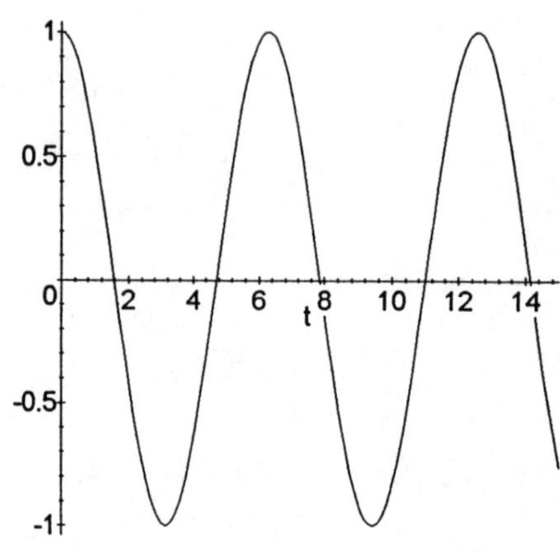

MATHEMATICA:

Task 1. Here is the 'spring' procedure. The variables are m for mass, d for damping, k for the spring constant, pos for the initial position, and vel for the initial velocity.   Type the following into Mathematica and save the procedure on a diskette for later use.

```
In[1]:=spring[m_,d_,k_,pos_,vel_]:=
 Block [ {},
   ms:=N[m]; dp:=N[d]; st:=N[k]; y0:=N[pos]; v0:=N[vel];
   springz[t_]:=y0 Exp[-dp t/2/ms] + (v0 + dp y0/2/ms) t Exp[-dp t/2/ms];
 springp[t_]:=Block[{r,s,r1,r2,c1,c2},
     r = -dp/2/ms; s = Sqrt[dp^2 - 4 ms st]/2/ms;
     r1 = r + s; r2 = r - s;
     c1 = (r2 y0 - v0)/(r2 - r1);
     c2 = (r1 y0 - v0)/(r1 - r2);
     c1 Exp[r1 t] + c2 Exp[r2 t]
    ];
   springn[t_]:= Block[{r,s,d1,d2},
    r = -dp/2/ms;
    s = Sqrt[4 ms st - dp^2]/2/ms;
    d1 = y0;
    d2 = (v0 - r y0)/s;
    Exp[r t] (d1 Cos[s t] + d2 Sin[s t])
    ];
   Which[dp^2 - 4 ms st == 0 , Plot[springz[t],{t,0,16}],
    dp^2 - 4 ms st > 0 , Plot [springp[t],{t,0,16}],
    dp^2 - 4 ms st < 0 ,Plot[springn[t],{t,0,25.2},
    Ticks -> {Range[0,25.2,Pi], Automatic}]
   ]
 ]
```

Task 2. To test the procedure, type:

```
spring[1,0,1,1,0];
```

Investigating the behavior of a mass-spring system: No damping

MAPLE:

Task 1. If necessary, retrieve the spring procedure and re-execute it. Test it again:

```
> spring(1,0,1,1,0);
```

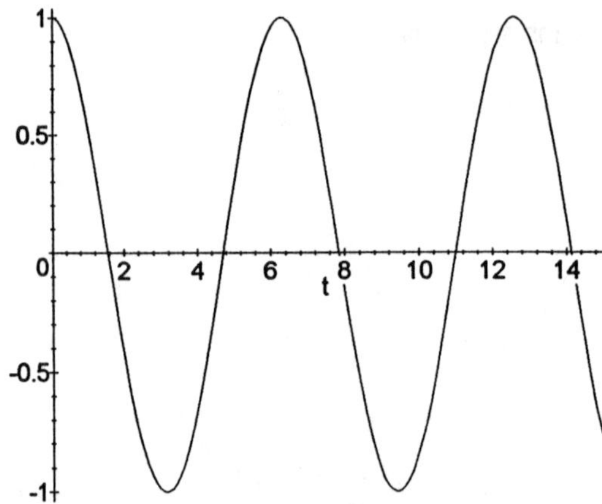

Tasks 2 and 3. Define a function of the parameter m and use it as follows:

```
> f:=m->spring(m,0,1,1,0);
```
$$f := m \rightarrow spring(m, 0, 1, 1, 0)$$

Here is the test graph again, using f :

```
> f(1);
```

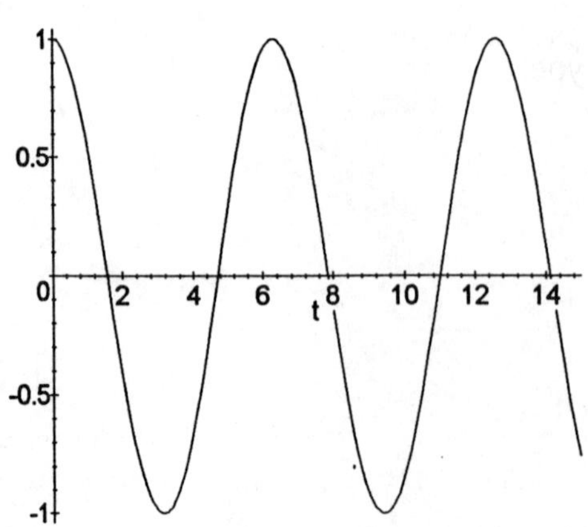

MATHEMATICA:

Task 1. If necessary, retrieve the spring procedure and re-execute it. Test it again:

In[1]:=spring[1,0,1,1,0];

Tasks 2 and 3. Define a function of the parameter m and use it as follows:

In[2]:=f[m_]:=spring[m,0,1,1,0];

Here is the test graph again, using f:

In[3]:=f[1];

Solving initial value problems, with and without your CAS

MAPLE:

Task 1. Using dsolve to find the general solution of a second order DE.

Example 1. We'll do Problem 1 of Task 1.

```
> eqn:=diff(y(t),t$2)-6*diff(y(t),t)+8*y(t)=0;#Define the equation
```

$$eqn := \left(\frac{\partial^2}{\partial t^2}\, y(t)\right) - 6\left(\frac{\partial}{\partial t}\, y(t)\right) + 8\, y(t) = 0$$

```
> gensol:=dsolve(eqn,y(t)); # Solve it
```

$$gensol := y(t) = \_C1\, e^{(2\,t)} + \_C2\, e^{(4\,t)}$$

Example 2. Here's an equation with decimal coefficients.

```
> eqn:=diff(y(t),t$2)-0.5*diff(y(t),t)-Pi*y(t)=0; # Another
  equation, with decimal numbers
```

$$eqn := \left(\frac{\partial^2}{\partial t^2}\, y(t)\right) - .5\left(\frac{\partial}{\partial t}\, y(t)\right) - \pi\, y(t) = 0$$

```
> gensol:=dsolve(eqn,y(t)); # Solve it
Error, (in factor/factor) floats not handled
```

Maple doesn't like the 0.5 coefficient. Let's change it to 1/2 :

```
> eqn:=diff(y(t),t$2)-(1/2)*diff(y(t),t)-Pi*y(t)=0;
```

$$eqn := \left(\frac{\partial^2}{\partial t^2}\, y(t)\right) - \frac{1}{2}\left(\frac{\partial}{\partial t}\, y(t)\right) - \pi\, y(t) = 0$$

```
> gensol:=dsolve(eqn,y(t)); # This gives the exact solution
```

$$gensol := y(t) = \_C1\, e^{(1/4(1+\sqrt{1+16\pi})\,t)} + \_C2\, e^{(-1/4(-1+\sqrt{1+16\pi})\,t)}$$

If you want approximations for the exponents, use the theory:

```
> chareqn:=r^2-.5*r-Pi=0; # The characteristic equation
```

$$chareqn := r^2 - .5\, r - \pi = 0$$

```
> rts:=fsolve(chareqn,r); # Solve it
```

$$rts := -1.539997948, \ 2.039997948$$

```
> y1:=exp(-1.54*t); # Build the first solution
```

$$y1 := e^{(-1.54\,t)}$$

```
> y2:=exp(2.04*t); # Build the second solution
```

$$y2 := e^{(2.04\,t)}$$

Let's test the first solution:

```
> test1:=subs(y(t)=y1,eqn);
```

$$test1 := \left(\frac{\partial^2}{\partial t^2}\,e^{(-1.54\,t)}\right) - \frac{1}{2}\left(\frac{\partial}{\partial t}\,e^{(-1.54\,t)}\right) - \pi\,e^{(-1.54\,t)} = 0$$

```
> simplify(test1);
```

$$.7346000000 \ 10^{-5}\,e^{(-1.540000000\,t)} = 0$$

The coefficient .7346*10^-5 is very close to zero, so y1 is a solution.

Example 3. Here's how Maple handles complex eigenvalues. This is Problem 3 of Task 1.

```
> eqn:=diff(y(t),t$2)-6*diff(y(t),t)+13*y(t)=0;
```

$$eqn := \left(\frac{\partial^2}{\partial t^2}\,y(t)\right) - 6\left(\frac{\partial}{\partial t}\,y(t)\right) + 13\,y(t) = 0$$

```
> gensol:=dsolve(eqn,y(t));
```

$$gensol := y(t) = \_C1\,e^{(3\,t)}\sin(2\,t) + \_C2\,e^{(3\,t)}\cos(2\,t)$$

Example 4. Let's try one with decimal coefficients:

```
> eqn:=diff(y(t),t$2)-(1/2)*diff(y(t),t)+Pi*y(t)=0;
```

$$eqn := \left(\frac{\partial^2}{\partial t^2}\,y(t)\right) - \frac{1}{2}\left(\frac{\partial}{\partial t}\,y(t)\right) + \pi\,y(t) = 0$$

```
> gensol:=dsolve(eqn,y(t));
```

$$gensol := y(t) = \_C1\,e^{(-1/4(-1+\sqrt{1-16\pi})t)} + \_C2\,e^{(1/4(1+\sqrt{1-16\pi})t)}$$

The answer is given in terms of complex-valued functions. If you want real solutions, use the theory:

```
> chareqn:=r^2-(1/2)*r+Pi=0; # The characteristic equation
```

$$chareqn := r^2 - \frac{1}{2}r + \pi = 0$$

```
> rts:=fsolve(chareqn,r,complex); # Solve it
```

$$rts := .2500000000 - 1.754734354\, I,\ .2500000000 + 1.754734354\, I$$

Now we'll pick off the two roots, get their real and imaginary parts, and build real solutions. We'll do r1 and let you do r2 .

```
> r1:=rts[1]; # Pick off the first root
```
$$r1 := .2500000000 - 1.754734354\, I$$
```
> rer1:=Re(r1); # Get the real part
```
$$rer1 := .2500000000$$
```
> imr1:=Im(r1); # Get the imaginary part
```
$$imr1 := -1.754734354$$
```
> y1:=exp(rer1*t)*cos(imr1*t); # Build a solution
```
$$y1 := e^{(.2500000000\, t)}\, \cos(1.754734354\, t)$$

We'll test this solution:

```
> test1:=subs(y(t)=y1,eqn);
```
$$test1 := \left(\frac{\partial^2}{\partial t^2}\, e^{(.2500000000\, t)}\, \cos(1.754734354\, t)\right) - \frac{1}{2}\left(\frac{\partial}{\partial t}\, e^{(.2500000000\, t)}\, \cos(1.754734354\, t)\right)$$
$$+\ \pi\, e^{(.2500000000\, t)}\, \cos(1.754734354\, t) = 0$$
```
> simplify(test1);
```
$$.1000000000\ 10^{-8}\, e^{(.2500000000\, t)}\, \cos(1.754734354\, t) = 0$$

It looks good.

Task 2. Solving an IVP with dsolve. We'll use the IVP of the example in the Lesson.

```
> eqn:=diff(y(t),t$2)-6*diff(y(t),t)+8*y(t)=0;
```
$$eqn := \left(\frac{\partial^2}{\partial t^2}\, y(t)\right) - 6\left(\frac{\partial}{\partial t}\, y(t)\right) + 8\, y(t) = 0$$
```
> sol:=dsolve({eqn,y(0)=1,D(y)(0)=3},y(t));
```
$$sol := y(t) = \frac{1}{2}\, e^{(2\, t)} + \frac{1}{2}\, e^{(4\, t)}$$

Let's check it:

```
> y1:=rhs(sol); # Capture the solution
```
$$y1 := \frac{1}{2}\, e^{(2\, t)} + \frac{1}{2}\, e^{(4\, t)}$$
```
> test1:=subs(y(t)=y1,eqn); #  Check that it's a solution
```

$$test1 := \left(\frac{\partial^2}{\partial t^2} \left(\frac{1}{2} e^{(2\,t)} + \frac{1}{2} e^{(4\,t)} \right) \right) - 6 \left(\frac{\partial}{\partial t} \left(\frac{1}{2} e^{(2\,t)} + \frac{1}{2} e^{(4\,t)} \right) \right) + 4\,e^{(2\,t)} + 4\,e^{(4\,t)} = 0$$

```
> simplify(test1);
```

$$0 = 0$$

Now check the initial conditions:

```
> eval(subs(t=0,y1));
```

$$1$$

```
> dy1:=diff(y1,t); # Find dy1/dt
```

$$dy1 := e^{(2\,t)} + 2\,e^{(4\,t)}$$

```
> evalf(subs(t=0,dy1));
```

$$3.$$

For completeness, we can plot this solution:

```
> plot(y1,t=-2..0.5);
```

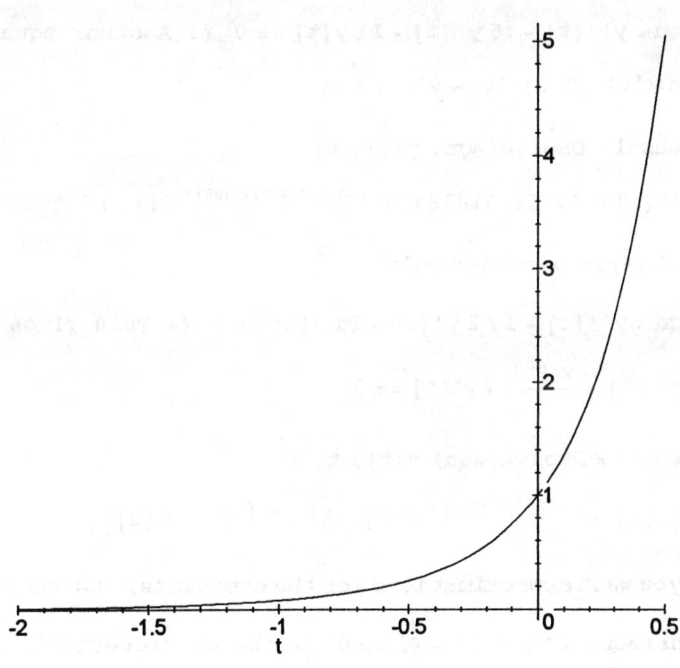

```
>
```

MATHEMATICA:

Task 1. Using DSolve to find the general solution of a second order DE.

Example 1. We'll do Problem 1 of Task 1.

```
In[69]:= eqn = y''[t] - 6 y'[t] + 8 y[t] == 0   (* Define the equation *)

Out[69]= 8 y[t] - 6 y'[t] + y''[t] == 0

In[70]:= gensol = DSolve[eqn, y[t], t]   (* Solve it *)

Out[70]= {{y[t] → E^(2t) C[1] + E^(4t) C[2]}}
```

Example 2. Here's an equation with decimal coefficients.

```
In[71]:= eqn = y''[t] - .5 y'[t] - Pi y[t] == 0  (* Another equation, with decimal numbers *)

Out[71]= -π y[t] - 0.5 y'[t] + y''[t] == 0

In[72]:= gensol = DSolve[eqn, y[t], t]
```

$$Out[72]= \{\{y[t] \to 2.71828182845905^{-1.53999794792893\,t}\,C[1] + 2.71828182845905^{2.03999794792893\,t}\,C[2]\}\}$$

... An approximation only.

```
In[73]:= eqn = y''[t] - 1/2 y'[t] - Pi y[t] == 0  (* This gives an exact solution *)
```

$$Out[73]= -\pi y[t] - \frac{y'[t]}{2} + y''[t] == 0$$

```
In[74]:= gensol = DSolve[eqn, y[t], t]
```

$$Out[74]= \left\{\left\{y[t] \to E^{\frac{1}{4}(1-\sqrt{1+16\pi})\,t}\,C[1] + E^{\frac{1}{4}(1+\sqrt{1+16\pi})\,t}\,C[2]\right\}\right\}$$

If you want approximations for the exponents, use the theory :

```
In[76]:= chareqn = r^2 - .5 r - Pi == 0  (* The characteristic equation *)

Out[76]= -π - 0.5 r + r^2 == 0

In[77]:= charsols = N[Solve[chareqn, r], 5]  (* This gives 5-place accuracy *)

Out[77]= {{r → -1.54}, {r → 2.04}}
```

In[78]:= y1[t_] := Exp[-1.54 t] (* Build the first solution *)

In[79]:= y2[t_] := Exp[2.04 t] (* Build the second solution *)

In[80]:= eqn /. y[t] -> y1[t] (* The wrong way to test the first solution *)

Out[80]= $-E^{-1.54t} \pi - \dfrac{y'[t]}{2} + y''[t] == 0$

In[81]:= eqn /. y -> y1 (* The correct way to test the first solution *)

Out[81]= $3.1416 \, E^{-1.54t} - E^{-1.54t} \pi == 0$

Example 3. Here's how Mathematica handles complex eigenvalues. This is Problem 3

In[82]:= eqn = y''[t] - 6 y'[t] + 13 y[t] == 0   (* A DE with complex eigenvalues *)

Out[82]= $13 y[t] - 6 y'[t] + y''[t] == 0$

In[83]:= gensol = DSolve[eqn, y[t], t]

Out[83]= $\{\{y[t] \rightarrow E^{3t} C[2] \cos[2 t] - E^{3t} C[1] \sin[2 t]\}\}$

Example 4. Let's try one with decimal coefficients:

In[84]:= eqn = y''[t] - .5 y'[t] + Pi y[t] == 0 (* What about decimal coefficients? *)

Out[84]= $\pi y[t] - 0.5 y'[t] + y''[t] == 0$

In[85]:= gensol = DSolve[eqn, y[t], t]

Out[85]= $\{\{y[t] \rightarrow 2.71828182845905^{0.250000000000000\,t} C[2] \cos[1.75473435413734 t] -$
$1.00000000000000 \; 2.71828182845905^{0.250000000000000\,t} C[1] \sin[1.75473435413734 t]\}\}$

You can fix this expression in either of the ways mentioned above . Here' s the theory approach .

In[86]:= chareqn = r^2 - .5 r + Pi == 0   (* Define the characteristic equation *)

Out[86]= $\pi - 0.5 r + r^2 == 0$

In[87]:= charsols = N[Solve[chareqn, r], 5] (* Solve it *)

Out[87]= $\{\{r \rightarrow 0.25 - 1.7547 I\}, \{r \rightarrow 0.25 + 1.7547 I\}\}$

In[88]:= **r1 = r /. First[charsols]** (* Pick off the two roots *)

Out[88]= $0.25 - 1.7547\,I$

In[89]:= **r2 = r /. Last[charsols]**

Out[89]= $0.25 + 1.7547\,I$

In[90]:= **rer1 = Re[r1]** (* Get the real part of r1 *)

Out[90]= 0.25

In[91]:= **imr1 = Im[r1]** (* Get the imaginary part of r1 *)

Out[91]= -1.7547

In[92]:= **y1[t_] := Exp[rer1 t] Cos[imr1 t]** (* Define one solution *)

In[93]:= **eqn /. y -> y1** (* Test the solution *)

Out[93]= $-3.01659\,E^{0.25\,t} \mathrm{Cos}[1.75473\,t] + E^{0.25\,t}\,\pi\,\mathrm{Cos}[1.75473\,t] - 0.877367\,E^{0.25\,t}\,\mathrm{Sin}[1.75473\,t] - 0.5\,(0.25\,E^{0.25\,t}\,\mathrm{Cos}[1.75473\,t] - 1.75473\,E^{0.25\,t}\,\mathrm{Sin}[1.75473\,t]) ==$
0

In[94]:= **Simplify[%]**

Out[94]= True

Task 2. Solving an IVP with dsolve. We'll use the IVP of the example in the Lesson.

In[96]:= **eqn = y''[t] - 6 y'[t] + 8 y[t] == 0**

Out[96]= $8\,y[t] - 6\,y'[t] + y''[t] == 0$

In[97]:= **sol = DSolve[{eqn, y[0] == 1, y'[0] == 3}, y[t], t]**

Out[97]= $\left\{\left\{y[t] \to E^{2t}\left(\dfrac{1}{2} + \dfrac{E^{2t}}{2}\right)\right\}\right\}$

In[98]:= **y1[t_] = y[t] /. First[sol]** (* Capture the solution *)

Out[98]= $E^{2t}\left(\dfrac{1}{2} + \dfrac{E^{2t}}{2}\right)$

A 27 - 7

In[99]:= **eqn /. y -> y1** (* Check that it is the solution *)

Out[99]= $6 E^{4t} + 12 E^{2t} \left(\dfrac{1}{2} + \dfrac{E^{2t}}{2} \right) - 6 \left(E^{4t} + 2 E^{2t} \left(\dfrac{1}{2} + \dfrac{E^{2t}}{2} \right) \right) == 0$

In[100]:= **Simplify[%]**

Out[100]= True

In[101]:= **y1[0]**

Out[101]= 1

In[102]:= **y1'[0]**

Out[102]= 3

In[103]:= **Plot[y1[t], {t, -2, .5}];** (* Plot the solution *)

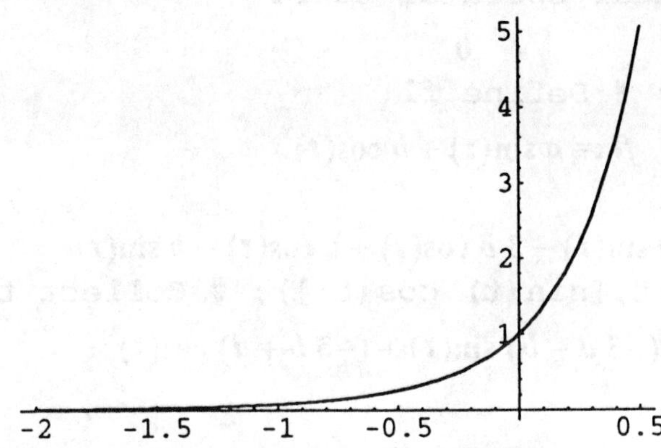

Linear operators

MAPLE:

Task 1. Define a linear operator as a mapping, then define functions to plug into the operator.

```
> L:=y->diff(y,t$2)+diff(y,t)-2*y; # Define the operator L as a
  mapping
```

$$L := y \rightarrow \text{diff}(y, t \, \$ \, 2) + \left(\frac{\partial}{\partial t} y \right) - 2\, y$$

```
> fa:=exp(t); # Define the function  fa  as an expression
```

$$fa := e^t$$

```
> L(fa); # Apply the linear operator to fa
```

$$0$$

```
> f1:=a*sin(t)+b*cos(t); # Define f1
```

$$f1 := a \sin(t) + b \cos(t)$$

```
> result:=L(f1);
```

$$result := -3\, a \sin(t) - 3\, b \cos(t) + a \cos(t) - b \sin(t)$$

```
> result2:=collect(result,[sin(t),cos(t)]); # Collect terms
```

$$result2 := (-3\, a - b) \sin(t) + (-3\, b + a) \cos(t)$$

MATHEMATICA:

Task 1. Define a linear operator as a mapping, then define functions to plug into the operator.

```
In[110]:=L[y_,t_]:=y"[t]+y'[t]-2 y[t]
```

```
In[105]:=fa[t_]:=Exp[t]
```

```
In[111]:=L[fa,t]
```

```
In[112]:=Clear[a,b]
```

```
In[113]:=fl[t_]:=a Sin[t]+b Cos[t]
```

```
In[114]:=L[fl,t]
```

```
In[115]:=Simplify[%]
```

Solving a nonhomogeneous equation: Undetermined coefficients

MAPLE:

__Task 3.__ We'll use problem 4 to show how to use Maple to do a lot of this algebra.

```
> restart; # Reset variables if you need to
> L:=y->diff(y,t$2)+diff(y,t)-2*y; # Define the linear operator
```

$$L := y \rightarrow \mathrm{diff}(y, t \,\$\, 2) + \left(\frac{\partial}{\partial t} y \right) - 2\, y$$

```
> trial:=A*cos(3*t)+B*sin(3*t); # Our trial solution
```

$$trial := A \cos(3\,t) + B \sin(3\,t)$$

```
> result:=L(trial); #Plug trial solution into the operator
```

$$result := -11\, A \cos(3\,t) - 11\, B \sin(3\,t) - 3\, A \sin(3\,t) + 3\, B \cos(3\,t)$$

```
> result2:=collect(result,[sin(3*t),cos(3*t)]); # Clean up the
  result
```

$$result2 := (-11\, B - 3\, A) \sin(3\,t) + (-11\, A + 3\, B) \cos(3\,t)$$

```
> sys:={-11*B-3*A=-1, -11*A+3*B=2}; # Define the system of equations
```

$$sys := \{-11\, B - 3\, A = \text{-}1,\ -11\, A + 3\, B = 2\}$$

```
> ans:=solve(sys,{A,B});
```

$$ans := \{A = \frac{\text{-}19}{130},\ B = \frac{17}{130}\}$$

```
> A:=rhs(ans[1]); # Pick off the solutions as coefficients
```

$$A := \frac{\text{-}19}{130}$$

```
> B:=rhs(ans[2]);
```

$$B := \frac{17}{130}$$

```
> trial; # Now Maple knows the solution
```

$$-\frac{19}{130} \cos(3\,t) + \frac{17}{130} \sin(3\,t)$$

```
> L(trial); # Check the solution in the operator
```

$$-\sin(3\,t) + 2 \cos(3\,t)$$

MATHEMATICA:

Task 3. We'll use problem 4 to show how to use Mathematica to do a lot of this algebra.

```
In[130]:=L[y_,t_]:=y''[t]+y'[t]-2 y[t]   (* Define the operator *)

In[131]:=Clear[a,b]   (* Clear undetermined coefficients *)

In[132]:=trial[t_]:=a Cos[3 t]+b Sin[3 t]   (* Define the trial solution *)

In[133]:=L[trial,t]   (* Plug the trial into the operator *)

In[134]:=Simplify[%]   (* Clean up the result *)

In[135]:=sys={-11 a+3 b==2,-3 a-11 b==-1}  (* Define the system of equations *)

In[136]:=ans=Solve[sys,{a,b}]   (* Solve it *)

In[137]:=a=a/.First[ans]   (* Pick off the solutions as coefficients *)

In[138]:=b=b/.Last[ans]

In[139]:=trial[t]   (* Now Mathematica knows the solution *)

In[141]:=Simplify[L[trial,t]]   (* Check the solution in the operator *)
```

The wronskian

MAPLE:

__Task 1__. Let's find the wronskian for Problem 1.

```
> with(linalg):  # Load the linear algebra package
Warning, new definition for norm
Warning, new definition for trace
> A:=vector([sin(t),cos(t)]); # Define the functions
```

$$A := [\sin(t), \cos(t)]$$

```
> Wr:=wronskian(A,t); # the wronskian matrix
```

$$Wr := \begin{bmatrix} \sin(t) & \cos(t) \\ \cos(t) & -\sin(t) \end{bmatrix}$$

```
> ans:=det(Wr);
```

$$ans := -\sin(t)^2 - \cos(t)^2$$

MATHEMATICA:

__Task 1__. Let's find the wronskian for Problem 1.

```
sys[t_]:={Cos[t],Sin[t]} (* Define the system of functions *)

w[t_]:={sys[t],sys'[t]} (* Define the matrix of functions and derivatives *)

MatrixForm[w[t]]  (* View it in matrix form *)

wro[t_]:=Det[w[t]] (* Define the Wronskian *)

wro[t]  (* View the Wronskian *)
```

Two nonlinear systems: Slope fields and phase portraits

MAPLE:

Task 1. Problems 3 and 4. We'll draw the slope field and the indicated trajectories with a single command. Execution of the phase portrait takes about a minute or two, so be patient when you use the command. There are many options available with the DEplot command, so you might want to look at the help screen by typing ?DEtools[DEplot] . Also, you might have to try several t-ranges before you get a satisfactory picture. Each problem is different.

```
> with(DEtools): # Load the DEtools package
> eqns:={diff(x(t),t)=3*x(t)-x(t)*y(t),diff(y(t),t)=-3*y(t)+x(t)*y(t
  )}; # Define the system as a set of equations
```

$$eqns := \{\frac{\partial}{\partial t}x(t) = 3\,x(t) - x(t)\,y(t), \frac{\partial}{\partial t}y(t) = -3\,y(t) + x(t)\,y(t)\}$$

```
> ic:={[x(0)=2,y(0)=3],[x(0)=2,y(0)=5],[x(0)=-1,y(0)=1],[x(0)=-1,y(0
  )=-1],[x(0)=1,y(0)=-1]}; # The initial conditions
```

$$ic := \{[x(0) = 2, y(0) = 3], [x(0) = 2, y(0) = 5], [x(0) = -1, y(0) = 1], [x(0) = -1, y(0) = -1],$$
$$[x(0) = 1, y(0) = -1]\}$$

```
> DEplot(eqns,{x(t),y(t)},t=0..10,ic,x=-5..5,y=-5..5,stepsize=.2);
```

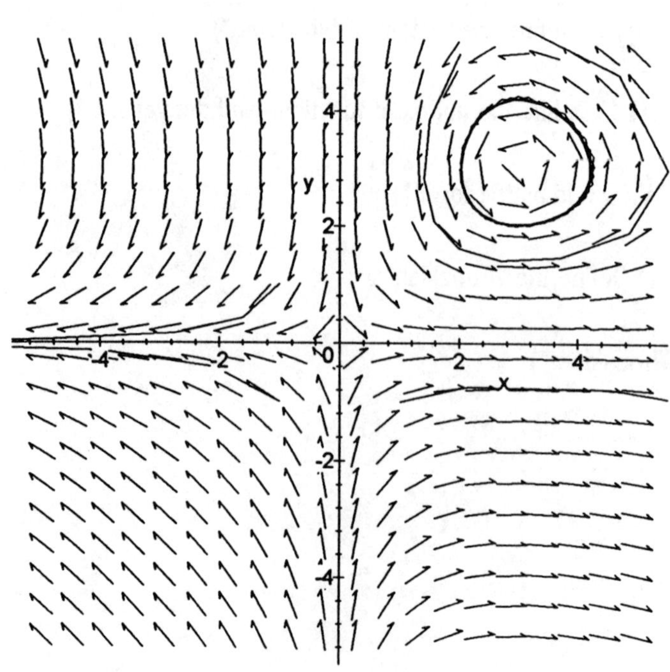

MATHEMATICA:

Task 1. Problems 3 and 4. First we'll draw the five trajectories of Problem 4, then the slope field, then plot them both on the same axes.

Define the system of DE's. Note the general initial conditions.

```
In[1]:=eqns[x0_,y0_]:={x'[t]==3 x[t]-x[t] y[t], y'[t]==-3 y[t]+x[t] y[t], x[0]==x0, y[0]==y0}
```

Now create the numerical solution to the system, again with no initial conditions yet specified.

```
In[2]:=sollist[t_,x0_,y0_]:={x[t],y[t]}/.NDSolve[eqns[x0,y0],{x[t],y[t]},{t,0,10}]
```

Next, generate five trajectories in the phase plane cooresponding to the five different initial conditions. There will be error messages before the graph appears. Don't worry about them.

```
In[3]:=
ParametricPlot[
  Evaluate[{sollist[t,2,3], sollist[t,2,5], sollist[t,-1,1],
    sollist[t,-1,-1], sollist[t,1,-1]} ],{t,0,10},
  PlotRange\[Rule]{{-7,7},{-7,7}}]
```

The PlotVectorField function used earlier for slope fields will also draw the direction field for a system. Note that the variable t is omitted from the syntax.

```
In[4]:=<<Graphics`PlotField`
```

```
In[5]:=PlotVectorField[{3x-x y,-3y+x y},{x,-7,7},{y,-7,7},ScaleFunction->(1&), Ticks->None]
```

Here are both graphs together.

```
In[6]:=Show[%3,%5]
```

Eigenvalues and eigenvectors

MAPLE:

__Task1.__ Problems 1 and 3. First we'll find the general solution of system (3), then solve the IVP.

```
> sys:={diff(x(t),t)=x(t)+y(t),diff(y(t),t)=3*x(t)-y(t)}; # Note the
  comma between the equations.
```

$$sys := \{\frac{\partial}{\partial t}x(t) = x(t) + y(t), \frac{\partial}{\partial t}y(t) = 3\,x(t) - y(t)\}$$

```
> fcns:={x(t),y(t)}; # The functions for which to solve
```

$$fcns := \{x(t), y(t)\}$$

```
> gensol:=dsolve(sys,fcns); # Solve the system
```

$$gensol := \{x(t) = \frac{1}{4}\_C1\,e^{(-2\,t)} + \frac{3}{4}\_C1\,e^{(2\,t)} + \frac{1}{4}\_C2\,e^{(2\,t)} - \frac{1}{4}\_C2\,e^{(-2\,t)},$$

$$y(t) = \frac{3}{4}\_C1\,e^{(2\,t)} - \frac{3}{4}\_C1\,e^{(-2\,t)} + \frac{3}{4}\_C2\,e^{(-2\,t)} + \frac{1}{4}\_C2\,e^{(2\,t)}\}$$

```
> ic:={x(0)=1,y(0)=2}; # Initial conditions
```

$$ic := \{x(0) = 1, y(0) = 2\}$$

```
> ivp:=sys union ic; # Create one list with equations and initial
  conditions
```

$$ivp := \{\frac{\partial}{\partial t}x(t) = x(t) + y(t), \frac{\partial}{\partial t}y(t) = 3\,x(t) - y(t), x(0) = 1, y(0) = 2\}$$

```
> psoln:=dsolve(ivp,fcns); # Solve the IVP
```

$$psoln := \{x(t) = -\frac{1}{4}e^{(-2\,t)} + \frac{5}{4}e^{(2\,t)}, y(t) = \frac{5}{4}e^{(2\,t)} + \frac{3}{4}e^{(-2\,t)}\}$$

```
> xp:=rhs(psoln[1]); # Pick off x(t)
```

$$xp := -\frac{1}{4}e^{(-2\,t)} + \frac{5}{4}e^{(2\,t)}$$

```
> yp:=rhs(psoln[2]); # Pick off y(t)
```

$$yp := \frac{5}{4}e^{(2\,t)} + \frac{3}{4}e^{(-2\,t)}$$

```
> test1:=subs(x(t)=xp,y(t)=yp,sys); # Check that we have a solution
```

$$test1 := \{\frac{\partial}{\partial t}\left(-\frac{1}{4}e^{(-2\,t)} + \frac{5}{4}e^{(2\,t)}\right) = \frac{1}{2}e^{(-2\,t)} + \frac{5}{2}e^{(2\,t)}, \frac{\partial}{\partial t}\left(\frac{5}{4}e^{(2\,t)} + \frac{3}{4}e^{(-2\,t)}\right) = -\frac{3}{2}e^{(-2\,t)} + \frac{5}{2}e^{(2\,t)}\}$$

```
> simplify(test1);
```

$$\{\frac{1}{2}\,(e^{(-4\,t)}+5)\,e^{(2\,t)}=\frac{1}{2}\,(e^{(-4\,t)}+5)\,e^{(2\,t)},\;-\frac{1}{2}\,(3\,e^{(-4\,t)}-5)\,e^{(2\,t)}=-\frac{1}{2}\,(3\,e^{(-4\,t)}-5)\,e^{(2\,t)}\}$$

```
> evalf(subs(t=0,xp)); # Check the initial conditions
```
1.

```
> evalf(subs(t=0,yp));
```
2.

MATHEMATICA:

Task1. Problems 1 and 3. First we'll find the general solution of system (3), then solve the IVP.

In[1]:=sys={x'[t]==x[t]+y[t], y'[t]==3 x[t]-y[t]} (* Define the system. *)

In[2]:=fcns={x,y} (* Define the functions for which to solve. *)

In[3]:=gensol=DSolve[sys,fcns,t] (* Solve it *)

In[4]:=x1[t_]=First[x[t]/.%3] (* Pick off the solutions. *)

In[5]:=y1[t_]=First[y[t]/.%3]

In[6]:=ic={x[0]==1,y[0]==2} (* Define initial conditions. *)

In[7]:=ivp=Join[sys,ic] (* Define one list with equations and initial conditions. *)

In[8]:=psoln=DSolve[ivp,{x,y},t] (* Solve the IVP. *)

In[9]:=xp[t_]=First[x[t]/.%8] (* Pick off the individual solution functions. *)

In[10]:=yp[t_]=First[y[t]/.%8]

In[11]:=Simplify[xp'[t]==xp[t]+yp[t]] (* Test the solutions in the system. *)

In[12]:=Simplify[yp'[t]==3 xp[t]-yp[t]]

In[13]:=xp[0] (* Test the initial conditions. *)

In[14]:=yp[0]

Plot[{xp[t],yp[t]},{t,-2,2}] (* Time graphs of the solutions. *)

ParametricPlot[{xp[t],yp[t]},{t,-2,2}] (* Phase plane trajectory of the solution. *)

A 49-2

A first look at Laplace Transforms

MAPLE:

Tasks 1. and 2.

```
> with(inttrans): # Load the integral transform package

> laplace(sin(3*t),t,s); # Transform of sin(3t). Note the order of
  t  and  s
```

$$\frac{3}{s^2 + 9}$$

```
> invlaplace(3/(s^2+9),s,t); # Inverse transform of the previous
  expression.  The order of  s  and  t  is reversed
```

$$\sin(3\,t)$$

MATHEMATICA:

Tasks 1. and 2.

```
In[12]:=<<Calculus`LaplaceTransform`  (* Load the Laplace transform package *)

In[13]:=LaplaceTransform[Sin[3 t],t,s]  (* Find the Laplace transform of a function. Note the order of t and s
*)

In[14]:=InverseLaplaceTransform[%,s,t]  (* The inverse Laplace transform of the previous expression. The order of
s and t is reversed *)

In[15]:=LT:=LaplaceTransform  (* Assign shorter aliases *)

In[16]:=ILT:=InverseLaplaceTransform

In[17]:=LT[Exp[a t],t,s]  (* Using the aliases *)

In[18]:=ILT[1/s,s,t]
```

APPENDIX B

SOFTWARE AND WEB SITES

SOFTWARE AND WEB SITES

We thank Sam Kaplan of Bowdoin College for compiling the following list of web site addresses.

COMPUTER ALGEBRA SYSTEMS

Maple V
Waterloo Maple
160 Columbia Street West
Waterloo, Ontario, Canada N2L3L3
phone: (519) 747-2373
email: info@maplesoft.on.ca

Mathematica
Wolfram Research, Inc.
100 Trade Center Drive
Champaign, Illinois 61820-7237, USA
phone: 217-398-0700
fax: 217-398-0747
email: info@wri.com

ODE SOLVERS

PHASER for the PC
Differential and Difference Equations through Computer Experiments
by Huseyin Kocak
with a Supplementary Diskette Containing PHASER: an Animator/Simulator for Dynamical
        Systems
Springer-Verlag, 175 Fifth Avenue, New York, NY 10010

MacMath 9.0 for the Macintosh
A Dynamical Systems Software Package
by John H. Hubbard and Beverly H. West
Springer-Verlag, 175 Fifth Avenue, New York, NY 10010

Differential Systems 4.0 for the Macintosh
U'Betcha Publications, 554 Evans Road, Springfield, PA 19064-3523
phone: (610) 544-9257
email: hgollwit@mcs.drexel.edu

ODE ARCHITECT, CD-ROM
Interactive Multimedia Modeling, Differential Equation Solving
Barbara Holland, John Wiley & Sons, 605 Third Avenue, New York, NY 10158-0012

<u>PROJECTS</u>

math.bu.edu/odes/
>   The homepage for the BU Differential Equations Project.

<u>SEARCH ENGINE</u>

www.maths.usyd.edu.au:8000/MathSearch.html
>   MathSearch is a search engine for math related sites including over 90,000 documents.

<u>LINKS</u>

archives.math.utk.edu/CTM/FIFTH/Ricardo/paper.html
>   Resource page for teaching differential equations including books, articles, software and
>   websites.

<u>SOFTWARE AND SOFTWARE ADD-ONS</u>

math.rice.edu:80/~polking/
math.rice.edu:80/~polking/odesoft/dfpp.html
>   John Polking is the author of the MATLAB routines dfield5 and pplane5. He has also
>   published a lab manual on differential equations using MATLAB. These URLs are his
>   home page and his page that refers to dfield and pplane.

www.microsim.com
>   Microsim Corp., makers of a PSPICE (circuit analyzer and designer) front end and a
>   wonderful demo version for FREE.

awi.aw.com/orderdemo.html
>   Interactive Differential Equations (IDE) CD-ROM by Beverly West, et al.
>   Addison Wesley Interactive, One Jacob Way, Reading, MA 01867

www.maplesoft.com/_navigation/documents/threeframes.html?demos&%2e%2e/%2e%2e/demos/
documents/intropage.html
>   A demo version of Maple.

www.cybermath.com
> "Live Math on the Web!"

rough.stanford.edu/
> Matlab script for ecology theory.

java.developer.com/pages/Gamelan.educational.html
> Java program site with topical index.

ike.engr.washington.edu/mathwright/
> Mathwright library. Mathwright is a program which makes it easy to generate interactive
> modules.

INTERACTIVE MODULES ON THE WEB

www.math.psu.edu/melvin/phase/newphase.html
> Richard Melvin has written a simple phase plane plotter in Java to be used over the web. It
> can also be used to plot slope fields.

www.sci.wsu.edu/idea/
> IDEA is the NSF-supported project for Internet Differential Equations Activities based at
> Washington State.

www.integrals.com/
> Integrals is a free Mathematica-driven integrator and uses Mathematica syntax.

links.math.rpi.edu/mods/diffeq.html
> Project Links: Differential Equations page. Site of interactive modules developed by
> Boyce, et al.

www.math.montana.edu/~frankw/ccp/home.htm
> Connected curriculum project. Interactive website including information on how to extract
> data from the web.

PHILOSOPHY

www.math.duke.edu/faculty/smith/thinking
> David Smith's essay on teaching .

<u>OTHER RESOURCES</u>

aslan.math.hmc.edu/codee/home.html
> C*ODE*E is the NSF-supported Consortium for Ordinary Differential Equations Experiments. They publish a nice newsletter.

archives.math.utk.edu/mathbio/
> UTK Mathematical Life Sciences archive homepage including a keyword search.

www.vernier.com/cblcat.shtml
> Caculator based lab system homepage allowing a hand-held calculator to collect real-world data.

www.ma.hw.ac.uk/solitons/
> Solitons Home Page. Includes recreating Scott Russell's soliton.

Other sites related to solitons:

> vizlab.rutgers.edu/
> www.math.uga.edu/~kasman/SOLITONPICS/
> next5.davidson.edu/~wc/WaveHTML/node39.html
> drip.colorado.edu/~blair/research/solitons/soliton_interactions.html
> drip.colorado.edu/~blair/research/research.html
> www.sfu.ca/~renns/images/pinch1.html   (light bullets)
> clap.inria.fr:8000/publication/Ercim_News/enw22/KdV'95.html
> w3.lab.kdd.co.jp/kdd/info/soliton/
> epswww.epfl.ch/ene/ene_apr96_conte_text.html
> www.cms.uncwil.edu/~herman/soliton2.htm
> www.cms.uncwil.edu/~herman/soliton.htm

Sites related to the Tacoma Narrows bridge:

> www.portoftacoma.com/tacoma/county/narrowsbridge.html
> www.fen.bris.ac.uk/engmaths/research/nonlinear/tacoma/tacoma.html
> wavelet.cycu.edu.tw/~chang/engmath/tacoma.html
> www.wsdot.wa.gov/eesc/environmental/Bridge-WA-99.htm
> www.harbornet.com/pna/bridge.html

next5.davidson.edu/~wc/WaveHTML/node39.html
drip.colorado.edu/~blair/research/solitons/soliton_interactions.html
drip.colorado.edu/~blair/research/research.html
www.sfu.ca/~renns/images/pinch1.html   (light bullets)
clap.inria.fr:8000/publication/Ercim_News/enw22/KdV'95.html
w3.lab.kdd.co.jp/kdd/info/soliton/
epswww.epfl.ch/ene/ene_apr96_conte_text.html
www.cms.uncwil.edu/~herman/soliton2.htm
www.cms.uncwil.edu/~herman/soliton.htm

Sites related to the Tacoma Narrows bridge:

www.portoftacoma.com/tacoma/county/narrowsbridge.html
www.fen.bris.ac.uk/engmaths/research/nonlinear/tacoma/tacoma.html
wavelet.cycu.edu.tw/~chang/engmath/tacoma.html
www.wsdot.wa.gov/eesc/environmental/Bridge-WA-99.htm
www.harbornet.com/pna/bridge.html